Soil, Plant and Water Analysis in Horticulture

NIPA® GENX ELECTRONIC RESOURCES & SOLUTIONS P. LTD.
New Delhi-110 034

Soil, Plant and Water Analysis in Horticulture

Chittaranjan Sarangi, Ph.D.
Assistant Professor (Soil Science)
College of Horticulture (OUAT)
Chiplima, Odisha – 768026

NIPA® GENX ELECTRONIC RESOURCES & SOLUTIONS P. LTD.
New Delhi-110 034

NIPA® GENX ELECTRONIC RESOURCES & SOLUTIONS P. LTD.

101,103, Vikas Surya Plaza, CU Block
L.S.C. Market, Pitam Pura, New Delhi-110 034
Ph : +91 11 27341616, 27341717, 27341718
E-mail: newindiapublishingagency@gmail.com
web: www.nipabooks.com

For customer assistance, please contact
Phone: + 91-11-27 34 17 17
Fax: + 91-11- 27 34 16 16
E-Mail: feedbacks@nipabooks.com

ISBN: 978-93-95763-97-4

Composed and Designed by NIPA®.

Dedicated to

The Seekers of Knowledge

Preface

Agricultural, horticultural, forestry and fodder production etc. for the burgeoning population are ever increasing enterprises that impose heavy tasks on the management of natural resources. Short term profitability at the cost of sustainability disharmonises the natural forces and processes resulting in peril of the inhabitants of the earth. To this end an understanding with conviction of the principles of soil, plant and water analysis as applied to horticulture will enhance the realisation of management goals in harmony with nature.

Intended primarily to be a text book, it will help educate learners to follow those principles. Students, planners, faculties and any one interested in achieving those goals will find it necessarily useful.

We take responsibility for any mistake. The readers are requested to communicate any such mistakes which will be duly taken care of.

Chittaranjan Sarangi

Acknowledgement

I deeply acknowledge the work itself that it could come to an end.

The endeavour put by various authors from which we have drawn freely has motivated us to undertake such a task is fully acknowledged.

I am thankful to the students teaching whom was an enlightened experience that stimulated me to reach wider audience through this work.

I am very much thankful to M/S New India Publishing Agency, New Delhi who have made it possible to bring the work in book form for the benefit of the readers.

The co- operation by each and everyone extended is cordially acknowledged.

Chittaranjan Sarangi

Contents

1

Introduction

Horticultural plants present a kaleidoscopic scenario from small herbs to very big deep rooted perennial trees. The associated rhizospheres present variable conditions as to supply of nutrients, water and air. The principles governing their physical, chemical and biological properties along with their relationship with plants need to be understood clearly. The plants differ in their requirements at various stages of growth; hence their tissues require constant monitoring of nutrient status for better productivity and quality. As nutrient behavior in soils is governed by soil properties and environmental conditions, measurement of such properties is often required. These include pH, salinity, organic matter, $CaCO_3$ and texture in drier areas the presence of Na and gypsum ($CaSO_42H_2O$) is also of concern.The irrigation water is a source of dissolved salts which are also nutrients and in excess impede water and nutrient relations of soil and plant. It requires a thorough understanding of the principles underlying methods of collection, characterization of water, air and nutrient relations of soil-plant system, monitoring of plant nutrient status, quality of irrigation water and soil and water pollution for effective management of nutrients avoiding pollution.

Soil and plant analysis is a diagnostic instrument for soil fertility and basis for fertilizer recommendation; to know where and where not fertilizer is to be applied. Obtaining accurate and precise values has always being the basis of soil analysis.

The aims of soil and plant analysis are:

1) To satisfy the demand for soil classification data.
2) To generate information for management and improvement of the soil.
3) To evaluate soil fertility in order to recommend fertilizer.
4) To determine the ecological effect of some agricultural production and environmental pollution.

It is important to have a clear idea about the purpose of any soil analysis as this will help determine sampling technique, sample preparation methods, elements or fractions to be determined and the analytical techniques to be employed.

Phases of soil testing

1) **Sample collection:** This should be such that it reliably reflects the average status of a field for the parameter considered.
2) **Extraction or digestion and nutrient determination:** The reagent used and the procedures adopted should quantify all or a portion of the element in the soil which is related to availability to the plant i.e. it should be correlated with plant growth.
3) **Interpreting the analytical results:** The units of measurement should reliably indicate if a nutrient is deficient, adequate, or in excess.
4) **Fertilizer recommendation:** This is based upon the soil test calibrated for field conditions, and considers other factors such as yield target, crop nutrient requirement, management of the crop, soil type, and method of fertilizer application.

2

Soil Sampling - Importance, Methods of Sampling and Processing of Soil Samples

Introduction

The objective of soil sampling is to obtain reliable information about a particular soil. The value of laboratory data in soil studies largely depends on effective sampling. Soil sampling is one of the most important aspects in the determination of the properties of the soil. Care in preparation and analysis can overcome the damages of careless or inappropriate sampling in the field (Bates, 1993). Errors arising from sampling are mainly related to spatial variability of the soil. While keeping in mind financial constraints, sampling procedures should be adopted properly to minimize the influence of spatial variability and to maximize the accuracy of mean plot values for the measured soil variables. The obtained soil data should be representative for the whole observation plot, taking into consideration the spatial variability of the soil within the plot.

Importance of soil sampling

Soil sampling and testing are done to provide an estimate of the capacity of the soil to supply adequate amount of nutrients to meet the needs of growing crops and information to its nature and problem. Traditionally, the goal of soil sampling is to develop a representative estimate of a field. A few grams of soil are actually used for the laboratory analysis. That small amount must represent the entire area for which the recommendation is to be made. The samples should be collected in such a way that it would represent the area for which recommendation is to be made.

Types of sampling

1) Simple random sampling
2) Systematic sampling
3) Stratified sampling

Sampling units

Each sample should represent 2 ½ acres or less for best characterization of the variability within the field, to serve as a guide for variable-rate application of plant nutrients. Where field variability is low, larger sample areas are acceptable; where variability is high; more samples are needed to adequately represent the field. Each field is to be treated as a separate sampling unit. Further, fields with varying soil types, slope and past management practices such as fertilization, liming and cropping pattern need to be treated as separate sampling units.

Sampling depth

Sampling depth should be determined to represent the root zone that the plant will draw from, but should also be consistent with the sampling depth used in developing the calibration data set to be used for interpreting the soil tests.

Prescribed depth of soil sampling

Field crops	15-20 cm
Deep rooted crops	30-60 cm (Sampling at different depths or layers is ideal)
Forage or pasture crops	8-10 cm
For immobile nutrients (P, K, Ca and Mg)	Sampling at tillage depth
Nitrate, Sulfate	60 cm (when the biological activity is low)
Saline alkali soils	Salt crust should be sampled separately and the depth of sampling should be recorded

Sampling time

Fields used for crop production are best sampled at any time after harvest and before planting. Collect samples after harvest, usually three to six months before planting or three months of after fertilizer application. Doing so gives time to plan a liming and fertilization programme before the busy planting season. Sample whenever subnormal growth or plant discoloration occurs. For coastal soils, collect samples every two years or test one half of the land every year. Sandy soils lose nutrients quickly and become acidic when nitrogen is added. For mountain or variegated soils, collect samples every three years or test one third of land every year. Silt and clay loam soils do not lose nutrients as quickly as sandy soils. Do not sample immediately after fertilizer has been applied. Soils should be tested as often as necessary to determine the influence of cultural practices and crop production on soil chemical properties. Nitrogen results are usually the most variable. So take samples as close to planting time as possible. Nitrate –N concentrations should be determined on annual basis. Phosphorus and potassium should be tested every three to four years. To account for seasonal variations, soil samples should be collected at approximately same time each year.

Not to sample

Dead or back furrows, fencerows, old or new, old roadbeds, or near limestone gravel roads, terrace channels, wind breaks or snow fence lines, turn-rows, spill areas.

Sampling points

Uniform fields can be sampled in a simple random, stratified random or in a systematic pattern. The soil test result from these sampling plans, provides an estimate of the entire population of possible soil test results. As the number of cores increases, the error or chance of inaccurate estimate of the average soil test value decreases. With a simple random system, each soil core is selected separately, randomly of previously drawn units. A stratified random sample is taken from a field that has been divided into several subunits or quadrants from which simple cores are obtained. The systematic sample is a further progression is an attempt to ensure complete field coverage, similar to the change from the simple random to the stratified random. Sample cores are taken at regularly spaced intervals in all directions.

Materials required

Spade, Khurpi, Auger (screw or post-hole type),Core sampler, Soil testing tube (for wet soil), Sampling bags, Plastic basin or bucket, Plastic sheet and Information sheet etc.

Collection of soil samples from the field

Normally each field may be treated as a sampling unit. But two or more fields, which are similar in appearance, production and past-management practices may be grouped together into a single sampling unit. Samples should be collected separately from areas, which differ in soil colour or past-management practices such as liming, fertilization, cropping pattern etc. During collection of soil, dead furrows, old manure or lime piles, wet spots, areas near trees, manure pits, compost pits, near field bunds and irrigation channels must be avoided. The sampling should be done in a zigzag pattern across the field to get homogeneity. A wise collecting agent is one who collects samples in the presence of the owner or cultivator of the land who is the best judge in deciding which area of his farm should be sampled separately. Scrap away the surface litter and insert the sampling auger to plough depth (15 cm). Take at least 15 samples randomly distributed over each area and put them in a clean bucket. If a sampling auger is not available make a 'V' or 'U' shaped cut to a depth of 15 cm or to a required depth using a spade and remove 1.0 to 1.5 cm thick slice of soil from top to bottom of the exposed face of the 'V' or 'U' shaped cut and put in a clean bucket or basin and in similar manner collect the soil sample from all the spots. Thoroughly mix the soil samples taken from 15 or more spots of each area. Remove foreign bodies such as plant roots, stubbles, leaves, glass pieces, pebbles, stones or gravels. By quartering technique, discard all but ½ to 1 kg soil. Quartering technique is done by dividing the thoroughly mixed soil into four equal parts and discarding two opposite quarters. Remix the remaining two quarters and again divide into four equal parts and reject the opposite two. Repeat this procedure until about ½ to 1 kg of soil is left. Instead of quartering, compartmentalization method can be followed. For this, spread the soil on a clean hard surface and mark lines from both the sides and create number of compartments. Take a little quantity of soil from each compartment and put into a clean container. Repeat the process of collection until the required quantity is collected. Store the soil in a clean cloth bag or container with proper labeling for further analysis.

Collection of soil samples from a profile

After the profile has been exposed, clean one face of the pit carefully with a spade and note the succession and depth of each horizon. Prick the surface with a knife or edge of the spade to show up structure, colour and compactness. Describe the profile as per the standard terminologies. Use the Munsell colour chart for noting the colour and find out the texture by feel method. Collect samples from each horizon by holding a large basin at the bottom limit of the horizon while the soil above is loosened by a khurpi. Mix sample and transfer to a bag after labelling.

Preparation of soil sample for analysis

Steps involved in the preparation and processing of soil sample are Drying Grinding. Sieving, Storing. The soil sample received at the laboratory is air-dried in shade and spread on a sheet of paper after breaking large lumps, if present, with a wooden mallet. It is further ground by pounding with a wooden mallet in such a way that the aggregate particles are broken down to ultimate soil particles. The soil thus prepared is sieved through a sieve with round holes,

2 mm in diameter. The material on the sieve is again ground and sieved till all aggregate particles are fine enough to pass through and only stones and organic residues remain on the sieve. For micronutrients analysis nylon sieve should be used. Mix well the 'fine soil' got by sieving and store in a suitable bottle or container with one label on the outside and another inside the container.

Objectives of soil testing

1. Evaluation of fertility status of soil
2. Estimation of the available nutrients status of soil
3. Evaluation of the suitability of soil for laying garden
4. Determination of acidity, salinity and alkalinity problems and
5. Recommendation of the required amount of fertilizers, lime or gypsum based on soil test value.

For determination of organic carbon, powder and sieve the soil through 0.2 mm sieve. For micronutrient analysis, iron, brass, copper and zinc containers must be avoided for collection and storage of soil samples and plastic implements and sieve must be used for sieving.

Sub sampling for analysis

Empty the soil in the bottle on a clean thick sheet of paper and spread evenly with a sampling knife. Heap the soil into a cone by raising the four ends of the paper and again mix it well and spread evenly on the paper as before. Repeat the process 3 or 4 times to ensure uniformity and then spread evenly on the paper again finally. Now divide the soil into four equal quarters and take a small quantity of soil from various points in each quarter to get a representative sample for analysis.

References

Agriculture Handbook 60. U.S. Dept. of Agriculture. Washington D.C.

Hesse, P. R. 1972. A Textbook of Soil Chemical Analysis. Published by Chemical Publishing Company Ltd., New York. pp. 556.

http://ecourses.iasri.rec.in/

http://eusoils.jrc.ec.europa.eu/soil_sampling/index.html

http://www.epa.gov/oust/cat/mason.pdf

http://www.sci.sdsu.edu/SERG/techniques/mfps.html

http://www.spectrumanalytic.com/services/analysis/plantguide.pdf

Jackson, M.L. 1973. Soil Chemical Analysis. Prentice Hall of India. New Delhi. pp. 498

Piper, C. S. 1942. Soil and Plant Analysis, Inter Science Publishers, Inc., New York

Soil Education - http://soils.usda.gov/education/index.html

Soil Science Education Home Page - http://soil.gsfc.nasa.gov

3

Plant Sampling; Importance, Methods of Sampling and Processing of Plant Samples

Introduction

Plant analysis can be defined as the quantitative determination of the concentration of an element or extractable fraction of an element in a sample from a particular part or portion of a crop. From the nutritional standpoint, plant analysis is based on the principle that the concentration of a nutrient within the plant is an integral value of all the factors that have interacted to it. Plant analysis is used as an index of available nutrient element supply. Plant growth or yield are compared with the elemental concentrations contained in the dry matter of the entire plant or plant structures such as leaves, petioles, fruit or grain sampled at different times during their development. Plant analysis gives the overall picture of the nutrient levels within the plant at the time the nutrient was taken. The use of plant analysis is based on the principle that the nutrient level present is as a result of all factors affecting the growth of the plant.

Plant analysis involves the determination of nutrient concentration in diagnostic plant part(s) sampled at recommended growth stage(s) of the crop. In a way plant analysis complements soil analysis. There are reliable sampling criteria and procedures for most of the world's commercial crops.

Importance

Plant analysis assesses nutrient uptake while soil testing predicts nutrient availability. The two tests are complementary to each other as crop management tools. Plant analysis will detect unseen hidden hunger and confirm visual deficiency symptoms. Toxic levels may also be detected. If it is done early, plant analysis will allow a corrective fertilizer application in the same season.

Sampling and analyzing tissue samples

A. Factors to be considered before sample collection

i. Nutrient element heterogeneity
ii. Statistical considerations

B. Sampling techniques

Factors on which the number of plants to sample are dependent

i. General condition of the plants
ii. Soil homogeneity
iii. Purpose for sampling

Methods of plant sampling

A basic knowledge of plant structure is necessary before collecting samples. A leaf is made up of a 'leaf blade' and a 'petiole'. The petiole is the stalk attached to the blade.

A compound leaf may have several 'leaflets' attached to it. In some cases, only terminal 'leaflets' may be sampled, as in the case of walnuts and pistachios. A common error in tomatoes is when only leaflets are sampled instead of the whole compound leaf. This shows the importance of understanding proper sampling technique.

The most recent mature leaf (MRML) is the first fully expanded leaf below the growing point. It is neither dull from age nor shiny green from immaturity. For some crops, the most recent mature leaf is a compound leaf. The most recent mature leaf on soybean and strawberry, for example, is a trifoliate compound leaf: three leaflets comprising one leaf. For cotton, grape, potato and strawberry, petioles provide an additional indication of nitrogen status. When sampling these crops, collect most recent mature leaves and their petioles. Detach leaves from petioles in the field to stop the translocation of nutrients. Put petioles in a separate bag. "Midribs" are the middle ribs to large leaves such as corn, lettuce, and cabbage, and would equate to a petiole sampling.

Deciding when to sample

To monitor plant nutrient status most effectively, sample during the recommended growth stages of the specific crop. Take samples weekly or biweekly during critical periods, depending on management intensity and crop value. However, to identify a specific plant growth problem, take samples whenever you suspect the problem.

The best time to collect samples is between mid-morning and mid-afternoon.

Nitrate nitrogen varies with time of day and prevailing conditions but generally not enough to alter interpretation. Sampling during damp conditions is okay but requires extra care to prevent tissue from decomposing during shipping. Keep samples free of soil and other contaminants that can alter results. The appropriate part of the plant to sample varies with the crop, stage of growth, and purpose of sampling. When sampling seedlings less than 4 inches tall, take whole plants from 1 inch above the soil line. For larger plants, the most recent mature leaf is the best indicator sample.

Taking a representative sample. Proper sampling is the key to reliable plant analysis results. A sample can represent the status of one plant or 20 acres of plants. In general, a common-sense approach works well. In problem solving, take samples from both 'good' and 'bad' areas. Comparison between the two groups of samples helps to pinpoint the limiting element. Comparative sampling also helps factor to out the influence of drought stress, disease, or injury. Take matching soil samples from the root zones of both 'good' and 'bad' plants for the most complete evaluation.

When monitoring the status of healthy plants, take samples from a uniform area.

If the entire field is uniform, one sample can represent a number of acres. If there are variations in soil type, topography, or crop history, take multiple samples so that each unique area is represented by its own sample.

Choosing sample size

The actual laboratory analysis requires less than one gram of tissue. However, a good sample contains enough leaves to represent the area sampled. Therefore, the larger the area is, the larger the sample size needs to be. Sample size also varies with the crop. For crops with large leaves like tobacco, a sample of three or four leaves is adequate. For crops with small leaves, like azalea, a sample of 25 to 30 leaves is more appropriate. For most crops, 8 to 15 leaves are adequate. For crops requiring petiole analysis, collect at least 15 to 20 leaves.

The table below shows the sampling guide for various crops

Crop	Time of Sampling	Where to Sample	No. of Plants to be sampled
Field Crops			
Alfalfa	Early bloom stage	Upper 1/3 of plant	12-30
Canola	Before seed set	Recently mature leaf	60-70
Clover	Before bloom	Upper 1/3 of plant	30-40
Corn/Sweet Corn	Seedling stage	All above ground portion	15-20
	Before heading	Upper 4 leaves	12-20
	Tasseling to silking	Opposite or below ear leaf	12-20
Grasses/forage mixes	Stage of best quality	Upper 4 leaves	30-40
Peanuts	Before / at bloom	Recently mature leaf	40-50
Small grains (barley, wheat, oat, rye, rice)	Seedling stage	All above ground portion	25-40
	Before heading	Upper 4 leaves	25-40
Sorghum	Before / at heading	2nd leaf from top	23-30
Soybeans	Before / at bloom	recently mature leaf	20-30
Sugarbeets	Midseason	recently mautre leaf at center of whorl	15-20
Sunflower	Before heading	Recently mature leaf	20-30
Tobacco	Before bloom	Recently mature leaf	10-15
Vegetable crops			
Asparagus	Maturity	Fern from 18 – 30 inches up	10-30
Beans	Seedling stage	All above ground portion	20-30
	Before / at bloom	Recently mature leaf	20-30
Broccoli	Before heading	Recently mature leaf	12-20
Brussel sprouts	Midseaason	Recently mature leaf	12-20
Celery	Midseaason	Outer petiole	12-20

In some specific cases, based on the nutrient to be analyzed, the sampling parts of the plant may vary. When leaves are sampled, recently matured ones are taken; both new and old growth is generally avoided. However, young emerging leaves are sampled for diagnosing iron chlorosis by determining ferrous (Fe^{++}) content of fresh leaves and B content in certain crops. Plant samples should be transported to the laboratory immediately in properly labeled paper. If samples are very wet, air-dry to a workable condition before packaging. Otherwise, decomposition or molding will occur. Include a completed plant analysis information sheet or cover letter with instructions within the same package.

Laboratory processing

Five steps are followed for processing the sampled plant tissues before analysis

1. Cleaning plant tissue to remove dust, pesticide and fertilizer residue normally by washing the plants with de-ionized water or with 0.1 to 0.3 % P-free detergents, followed by deionized (DI) water. If not essentially required,

samples for soluble element determination may not be washed, particularly for long periods. However, samples for total iron analysis must be washed. The washing procedure should be done as quickly as possible so as to prevent the loss of nutrients like K, Ca and Na through leaching. Immediate drying in an oven to stop enzymatic activity, usually at 65°C for 24 hours in a dust-free forced-draft electrical oven to a constant weight. This temperature satisfies the two separate requirements that must be met when drying plant material for analysis, they are: A sufficiently high temperature to destroy the enzymes responsible for decomposition processes. Optimum temperature for moisture removal without appreciable thermal decomposition Plant samples are therefore not recommended to be air dried because the temperature when air dried will not satisfy the above requirements.

2. Mechanical grinding to produce a material suitable for analysis, usually to pass a 60-mesh sieve; stainless mills are preferable, particularly when micronutrients analysis is involved.
3. Since most analytical methods require grinding of a dry sample, careful attention must be given to avoid contamination with the element to be analyzed. Particular care is required for micronutrients (metal grinder should not be used for micronutrients analysis). Equipments having grinding surfaces of either steel, stainless steel or agate are recommended. Examples of grinding equipments are: hammer mill, wiley mill, agate or glass mortar with pestle.
4. Final drying at 65°C of ground tissue to obtain a constant weight.
5. Storing in appropriate container. Dried ground plant material should not be stored in the shelf for more than two months to avoid becoming moldy. Samples to be kept for longer periods should be stored in a sterilized sealed bottle in a refrigerator at -5°C.

Moisture factor

Weighing of perfectly oven–dried samples is, however cumbersome (involves continuous oven-drying and use of desiccator, and is still prone to error) as plant material may absorb moisture during the weighing process, particularly if relative humidity is high in the laboratory. To get over this difficulty, use of the moisture factor is suggested instead. The moisture factor for each batch of samples can be calculated, by oven-drying only a few subsamples from the lot (e.g. 5 from a batch of 100-200 samples).

$$\text{Moisture factor} = \frac{\text{Weight of air dry sample (g)}}{\text{Weight of oven dry sample (g)}} \times 100 \qquad (1)$$

Thereafter, air dry samples are weighed, considering the moisture factor. For example, if moisture factor = 1.09, then weight of oven dry and air dry samples will be as follows.

Oven dry weight	Air dry weight
0.25	0.27
0.50	0.55
1.00	1.09
2.00	2.18

The moisture factor approach is also used for weighing soil samples, and expressing the analytical results on oven dry soil weight basis.

Plant analysis

Most methods developed for the analysis involve ashing the plant tissue to destroy the organic components leaving the various elements for analysis. Some scientists have proposed the use of extraction procedures which utilize dried, green tissue. There are two methods of ashing plant samples:

Wet ashing: Plant tissue is decomposed by digesting the plant sample in strong acid solutions such as HNO_3, H_2SO_4 and $HclO_4$. (Special precautions against explosions must be taken when $HclO_4$ is used). Wet digestion procedure is to be preferred for analysis of trace elements and elements which could be lost by volatilization in the dry ashing procedure (N, Cl, S). Since boron can be lost by steam distillation as H_3BO_3 and can be volatilized during wet ashing, the wet digestion procedure is not applicable for analysis of this element. There is the danger of loss of P by volatilization as H_3PO_4 during $HclO_4$ digestion if the temperature exceeds 230°C.

Dry ashing: Dry ashing consists of heating the sample to a temperature sufficiently high to burn off the carbon. This is normally carried out in a muffle furnace at temperatures not exceeding 500°C. Ashing may require two or more hours depending on the type of tissue. For highly carbonaceous tissues a longer ashing period may be necessary. The use of high walled crucibles are preferred over open flat vessels. Boron can be determined only by dry ashing since this element is volatilized during wet ashing. Heating of sample should not be too rapid during dry ashing, if heating is too rapid, the sample might burst into flames and excessively high temperatures will be reached. High temperatures are to be avoided because of the following reasons: Elements like Na, K, P, Cl, S and B might be lost as volatile compounds. No loss of P is expected below 600°C.

Salts present in the ash might fuse and form a fused mass surrounding particles of incompletely burned tissue. Complex silicates might form as siliceous residues in the ashed material. Such residues frequently contain significant amounts of trace elements and possibly some Na and K.

Method of analysis

- The following are the commonly used methods for the analysis of plant tissue ash
- Colorimetric
- Flame emission Most of the elements contained in plant tissue are present as constituents of the plant tissue rather than as water soluble inorganic anions or cations. Consequently, organic matter of plant tissue must be destroyed before the mineral elements can be determined.

i. Methods of organic matter destruction

   a. Wet Ashing- Decomposition of plant tissue by digesting in strong acid solutions

   b. Dry Ashing- Heating plant samples to a temperature sufficiently high to burn off the carbon

ii. Methods of determining elements in plant samples

- Total Nitrogen

Method: Micro-Kjeldahl

- Phosphorus

Method: Vanado-Molybdate

- Potassium

Method: Flame emission

- Calcium, magnesium, manganese, zinc iron and copper determination

Method: Atomic absorption

References

Garn A. Wallace. 2010. Plant Tissue Analysis Complements Soil TEST. www.wlabs.com copyright
http://www.avocadosource.com/books/TraynorJoe1980/IDEAS_PG_79-84.pdf
Illustrated Guide to Sampling for Plant Analysis.2009.Spectrum Analytic Inc
PIPER, C. S. 1942. Soil and Plant Analysis, Inter Science Publishers, Inc., New York
www.bettersoils.com/planttestingart.pdf
www.servitechlabs.com/Portals/0/PlantSampling1.pdf

4

Hydraulic Conductivity, Darcy's Law and eTermination of Hydraulic Conductivity

Three types of water movement within the soil are recognized. They are

i) Saturated flow
ii) Unsaturated flow
iii) Water vapour movement

Saturated flow

This occurs when the soil pores are completely filled with water. This water moves at water potentials larger than 33 kPa. Saturated flow is water flow caused by gravity's pull. It begins with infiltration, which is water movement into soil when rain or irrigation water is on the soil surface. When the soil profile is wetted, the movement of more water flowing through the wetted soil is termed percolation.

Hydraulic conductivity can be expressed mathematically as

$$V = kf$$

Where,

V = Total volume of water moved per unit time

f = Water moving force

k = Hydraulic conductivity of soil

Vertical water flow

Darcy's law gives the vertical water flow rate through soil. The law states that the rate of flow of liquid or flux through a porous medium is proportional to the hydraulic gradient in the direction of flow of the liquid. The quantity of water per unit time Q/t that flows through a column of saturated soil can be expressed by Darcy's law as follows

$$Q/t = -A\, K_{sat}.\Psi\Delta/L$$

Where A is the cross sectional area of the column through which the water flows, K_{sat} is the saturated hydraulic conductivity, Äø is the change in water potential between the ends of the column (for example $\Psi 1 - \Psi 2$) and L is the length of the column. For a given column, the rate of flow is determined by the ease with which the soil transmits water (K_{sat}) and the amount of force driving the water, namely the water potential gradient $\Delta\Psi/L$. For saturated flow this force may also be called as hydraulic gradient.

The units in which K_{sat} is measured are length/time, typically cm/s or cm/h. The K_{sat} is an important property that helps determine how well soil or soil material will perform in such uses as irrigated cropland, sanitary landfill cover material, waste water storage lagoon lining and septic tank drain field.

Values of saturated hydraulic conductivity and interpretation for soil uses

K_{sat}(cm/h)	Comments
36	Typically a beach sand
18	Typically a very sandy soil, too rapid to effectively filter pollutants in waste water
1.8	Typically a moderately permeable soil, K_{sat} between 1.0 and 15 cm/h considered suitable for most agricultural, recreational and urban uses calling for good drainage.
0.18	Typically a fine textured, compacted or poorly structured soils. Too slow for proper operation of septic tank drain fields, most types of irrigation and many recreational uses such as playgrounds.
< 3.6 x 10-5	Extremely slow, typical of compacted clay. K_{sat} of 10-5 to 10-8 cm/h may be required where nearly impermeable material is needed, as for wastewater lagoon lining or landfill cover material.

Factors influencing the hydraulic conductivity of saturated soils

Any factor affecting the size and configuration of soil pores will influence hydraulic conductivity. The total flow rate in soil pores is proportional to the fourth power of the radius. Thus, flow through a pore 1 mm in radius is equivalent to that in 10,000 pores with a radius of 0.1 mm even though it takes only 100 pores of radius 0.1mm to give the same cross-sectional area as a 1 mm pore. As a result, macropores (radius >0.05mm) account for most water movement in saturated soils. The presence of bio-pores, such as root channels and earthworm burrows (typically >1 mm in radius) may have a marked influence on the saturated hydraulic conductivity of different soil horizons.

Because they usually have more macro pore space, sandy soils generally have higher saturated hydraulic conductivities than fine-textured soils. Likewise, soils with stable granular structure conduct water much more rapidly than do those with unstable structural units, which break down upon being wetted. Saturated hydraulic conductivity of soils under natural vegetation is commonly much higher than in soils where cultivated crops have been grown. Entrapped air, which is common in recently wetted soils, can slow down the movement of water and thereby reduce the hydraulic conductivity.

Determination of saturated hydraulic conductivity

Saturated hydraulic conductivity (K_{sat}) of soil is generally determined in the laboratory on undisturbed soil core with the help of either constant head permeameter or falling head permeameter. In the first method, undisturbed soil core is collected by depth core sampler. After covering the bottom end with a muslin cloth, the soil core is saturated by placing it in a trough of water. The saturated soil core is then placed in the constant head permeameter where water is allowed to flow through it under a constant head maintained by a constant head reservoir. The steady state outflow through soil core is measured and used for calculating Ksat by Darcy's law:

$$K_{sat} = -VL/At\ \Delta H$$

Where, - sign indicates flow is from higher to lower potential, V is the outflow volume of water, L is the length of soil, A is the cross-sectional area of the soil, t is the time and ΔH is the pressure head difference inflow and outflow ends.

(ii) Unsaturated flow

It is flow of water held with water potentials lower than 1/3 bar. Water will move toward the region of lower potential (towards the greater "pulling" force). In a uniform soil this means that water moves from wetter to drier areas. The water movement may be in any direction .The rate of flow is greater as the water potential gradient (the difference in potential between wet and dry) increases and as the size of water filled pores also increases. The two forces responsible for this movement are the attraction of soil solids for water (adhesion) and capillarity. Under field conditions this movement occurs when the soil macropores (non-capillary pores) with filled with air and the micro pores (capillary pores) with water and partly with air.

Factors affecting the unsaturated flow

Unsaturated flow is also affected in a similar way to that of saturated flow. Amount of moisture in the soil affects the unsaturated flow. The higher the percentage of water in the moist soil, the greater is the suction gradient and the more rapid is the delivery.

(iii) Water vapour movement

The movement of water vapour from soils takes place in two ways-

(a) Internal movement—the change from the liquid to the vapour state takes place within the soil, that is, in the soil pores and

(b) External movement—the phenomenon occurs at the land surface and the resulting vapour is lost to the atmosphere by diffusion and convection.

The movement of water vapour through the diffusion mechanism takes place from one area to the other soil area depending on the vapour pressure gradient (moving force), this gradient being simply the difference in vapour pressure of two points a unit distance apart. The greater this difference, the more rapid the diffusion and the greater is the transfer of water vapour during a unit period.

Soil conditions affecting water vapour movement

There are mainly two soil conditions that affect the water vapour movement namely moisture regimes and thermal regimes. In addition to these, the various other factors which influence the moisture and thermal regimes of the soil are like organic matter, vegetative cover, soil colour etc. The movement takes place from moist soil having high vapour pressure to a dry soil (low vapour pressure). Similarly the movement takes place from warmer soil regions to cooler soil region. In dry soils some water movement takes place in the vapour form and such vapour movement has some practical implications in supplying water to drought resistant plants.

Entry of water into soil

Infiltration

Infiltration refers to the downward entry or movement of water into the soil surface. It is a surface characteristic and hence primarily influenced by the condition of the surface soil. Soil surface with vegetative cover has more infiltration rate than bare soil. Warm soils absorb more water than colder ones. Coarse surface texture, granular structure and high organic matter content in surface soil, all help to increase the infiltration. Infiltration rate is comparatively lower in wet soils than in dry soils.

Factors affecting infiltration: Clay minerals, Soil texture, Soil structure, Moisture content, Vegetative cover and Topography

Percolation

The movement of water through a column of soil is called percolation. It is important for two reasons.This is the only source of recharge of ground water which can be used through wells for irrigation. Percolating waters carry plant nutrients down and often out of reach of the plant roots (leaching). In dry region it is negligible and under high rainfall it is high.Sandy soils have greater percolation than clayey soil. Vegetation and high water-table reduce the percolation loss

Permeability

It indicates the relative ease of movement of water with in the soil. The characteristics that determine how fast air and water move through the soil is known as permeability. The term hydraulic conductivity is also used which refers to the readiness with which a soil transmits fluids through it.

References

Daniel Hillel. Fundamentals of Soil Physics. Academic Press.

Gildayal, B.P. and R.P. Tripathi. 1987. Soil Physics. Published by John Wiley & Sons (Asia) Pte Ltd., Singapore

Oswal, M.C. 1983. Soil Physics. Vikas Publishing House, New Delhi

Rattan Lal, R. Lal, Manoj Shukla.2004. Principles of Soil Physics. CRC press

T.D. Biswas and S.K. Mukherjee.1995.Text book of Soil Science. Tata McGraw-Hill Publishing Company Limited, New Delhi.

5

Renewal of Gases in Soil and Their Abundance

Soil air, its composition, gaseous exchange

Soil air is a continuation of the atmospheric air. It is in constant motion from the soil pores into the atmosphere and from the atmosphere into the pore space. The circulation of air in the soil and renewal of component gases like oxygen and carbon dioxide is known as soil aeration. Soil aeration is essential for the respiration and survival of soil organisms and plant roots.

Composition of soil air

Soil air contains gases like nitrogen, oxygen, carbon dioxide, water vapour and others. The composition of soil air is different from atmospheric air. Soil air contains more carbon dioxide and less oxygen than atmospheric air. It also contains more water vapour than atmospheric air. The nitrogen content of soil air is almost equal to atmospheric air. (Table below)

Source	Per cent by volume			
	Nitrogen	Oxygen	Carbon dioxide	Water vapour
Soil air	79.2	11.5-19	0.3-3	1.5-2
Atmospheric air	79.0	21.0	0.03	1-2

Soil aeration problems in the field

Factors affecting the composition of soil air

Nature of soil

Soil with more pore space volume will have more air. The drainage of excess water from soil macropores improves soil aeration. If the volumes of soil macropores are high, soil aeration will be good. Soil texture, bulk density, aggregate stability and organic matter contents influence the soil macropores. The quantity of oxygen in soil air is less than that in atmospheric air. The oxygen content of the air reduces with increasing depth of soil. Surface soil will have more oxygen than sub soil because of quick replenishment from the atmospheric air through diffusion. Light textured soil or sandy soil contains more air than the heavy soil. The concentration of CO_2 is more in subsoil due to slow rate of replenishment and low aeration in lower layer than in the surface soil. High clay containing soils will have high soil moisture potential - which reduces the oxygen diffusion rate and increases the CO_2 accumulation.

Type of crop

Plant roots require oxygen, which they take from the soil air and deplete the concentration of oxygen in the soil air and release CO_2. Soils on which crops are grown contain more CO_2 than fallow lands. The amount of CO_2 is more near the plant than farther away due to respiration by roots.

Microbial activity

The microorganisms in soil require oxygen for respiration and they take it from the soil air depleting its concentration. Decomposition of organic matter produces CO_2, because of increased microbial activity. Hence, soils rich in organic matter contain higher CO_2.

Seasonal variations

The oxygen content of soil air is higher in dry season than during the monsoon. During the dry season, most of the soil pores are filled with air and the exchange of gases between the soil air and the atmosphere is more. During monsoon seasons, most of the soil pores are filled with water. Temperature also influences the gaseous composition of the soil air.High temperature enhances the gaseous exchange between the soil air and the atmosphere. High temperature also increases the microbial respiration and releases more CO_2 in the soil air. Under field conditions, poor soil aeration occurs due to two conditions: a) when the moisture content is too high occupying most of the pore space and b) when the exchange of gases with the atmosphere is slow. When the field is completely submerged with water, all the plants except some plants like rice, die for want of oxygen.

Exchange of gases between soil and atmosphere

The exchange of gases between the soil air and the atmosphere is facilitated by two mechanisms.

(i) Mass flow (Convection)

During the rain fall or during irrigation, a part of the soil air is replaced by water and the replaced air moves out into the atmosphere.When the soil moisture is lost due to evaporation, plant absorption, internal drainage *etc.*, atmospheric air reenters into the soil pore space. The variations in soil temperature cause changes in the temperature of the soil air. Also, when the soil air gets heated during the day, it expands and the expanded air moves out into the atmosphere.When the soil cools during the night, the soil air contracts and the atmospheric air is drawn in.

It is modeled as

$$q_{air} = -(\kappa/\eta).VP$$

(where q_{air} is the (volume)flux of air, - sign is from higher pressure to lower pressure, k is the permeability of soil through the air filled prorsity, ç is the coefficient of viscosity of soil air, P is the pressure gradient in three directions)

In terms of mass of air

$$q_{am} = -(\rho\kappa/\eta).\ VP$$

(where q_{am} is the flux of mass of air, - sign is from higher pressure to lower pressure, ñ is the density of air, k is the permeability of soil through the air filled prorsity, ç is the coefficient of viscosity of soil air, P is the pressure gradient in three directions)

(ii) Diffusion

Most of the gaseous interchange in soils occurs by diffusion. Atmosphere and soil air contains gases such as nitrogen, oxygen, carbon dioxide etc., each of which exerts its own partial pressure in proportion to its concentration. These gases move from higher concentration (higher partial pressure) to lower concentration (lower partial pressure). Oxygen moves from the atmosphere into soil (air) and CO_2 moves out of the soil air into the atmosphere through diffusion without the movement of entire air mass because of the difference in their concentration or partial pressure. This movement is called diffusion and this will continue till equilibrium is established.

Theory of gas diffusion in the soil

Molecular gas diffusion in soils is controlled by the concentration gradient and the diffusion coefficient. Mathematically, the movement of any gas in the soil under steady state conditions occurs according to the following differential equation, which is Fick's first law:

$$J=Q/At=-D_f\,(dc/dx) \tag{1}$$

where J is the flux of a given gas, Q is the mass (g), A is the area (m^2), t is time, D_f is the diffusion coefficient of the gas in the soil, c is the concentration of the gas ($g\ m^{-3}$) and x is the distance. The processes that occur in the soil are more complex than those considered by Fick's First Law, since the flux of a gas in the soil varies over time; that is, gas transport does not occur under steady state conditions (Figure 1).

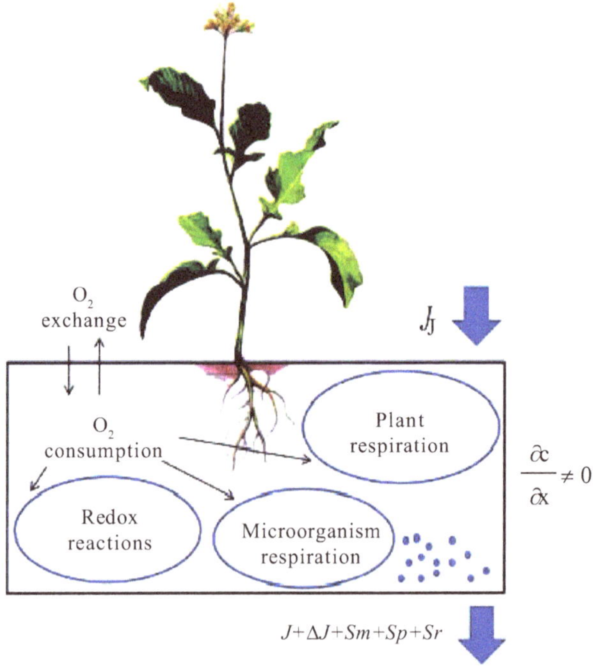

Fig. 1. Scheme of soil oxygen dynamics. J is gas density flux, c is gas concentration ($g\ m^{-3}$), t is time, x is distance, Sm is microorganism respiration, Sp is plant respiration, and Sr is redox reactions in the soil.

Thus, the movement of oxygen in the soil is described by an equation applicable to transient conditions which also considers oxygen sources and sinks.

$$\partial c/\partial t=-D_f\,(d^2c/dx^2)-S_g(S_{g,t}) \tag{2}$$

where c is gas concentration, t is time, x is distance and S_g represents the consumption and/or evolution of oxygen due to different sources (microorganisms, roots, oxidation-reduction processes) and depends upon time and distance. Equation [2] has the following assumptions: (1) Diffusion is the only mechanism for oxygen transport; convection is insignificant; and (2) macroscopic diffusion occurs in only one dimension.

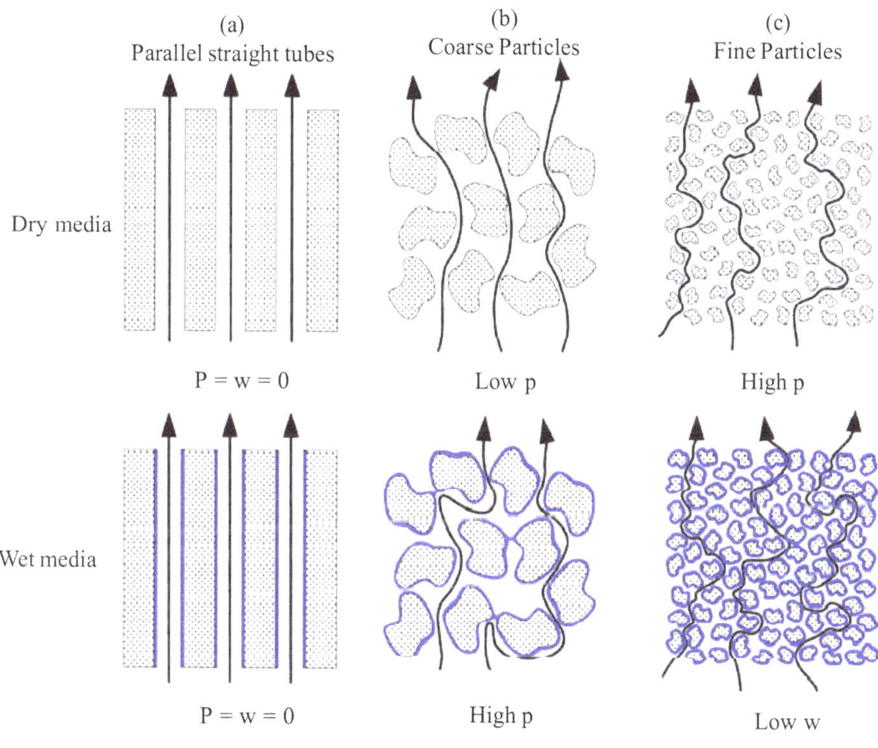

Source: Thorbjern *et al.* (2008)

Fig. 2: Gas diffusion in parallel straight tubes (a), coarse particles (b), and fine particles (c). The presence of single particles and the addition of water enhance the tortuosity and reduce the connectivity of the gas phase compared with straight parallel tubes, with the texture governing the resistance of the particles (reflected by the particle shape factor p) and the resistance of the water (reflected by the water-blockage factor w), respectively.

The soil gas diffusion coefficient D_p incorporating the effects of soil tortuosity and porosity; it is usually estimated by the following equation:

$$D_{p} = D_{0.} \tau \varepsilon \quad (3)$$

where D_0 is the diffusion coefficient of the gas in air ($m^2\ s^{-1}$), e is the soil air-filled porosity (m^3 air m^{-3} soil) and $\hat{o}$ is the tortuosity factor of the soil. The soil porosity filled with air (e) is a factor essentially controlled by the water-filled porosity (Φ) and the total porosity (Φ), which depend upon the soil type, organic matter, and management. The total soil porosity is mainly dependent on the texture and structure of the soil.

Importance of soil aeration

Soil reactions and properties

Microbial breakdown of soil organic residues is reduced under poor aeration and hence organic matter is accumulated. In well-aerated soil, aerobic microorganisms are active and they convert simple sugars to CO_2 and water using oxygen. In submerged or reduced soil, anaerobic microorganisms are active and they convert sugars to less CO_2 and more CH_4 (methane) which is an atmospheric pollutant. This process also gives out some organic acid, ethylene gas etc., which are toxic to plant roots and some microbes. The aerobic decomposition will be faster than anaerobic decomposition.

Oxidation and reduction of inorganic elements

The oxidized state of nitrogen and sulphur are easily utilized by the plants. Reduced forms of some of the elements are toxic. Though solubility of iron and manganese increases, they become toxic to the plants.

Soil colour is also altered by aeration. Well-aerated soils have red, yellow and reddish brown colours. Reduced soils have grey and blue colours. The discolouration of soil in patches is called mottling.

Plant and root growth

Soil aeration is important factor for the normal growth of plants. Roots absorb oxygen for their respiration and release CO_2. The supply of oxygen to roots in adequate quantities and the removal of CO_2 from the soil air are very essential for healthy plant growth. When the supply of oxygen is inadequate, the plant growth either slows or ceases completely as the accumulated CO_2 hampers the growth of plant roots. The abnormal effect of insufficient aeration on root development is most noticeable on the root crops. Abnormally shaped roots of these plants are common on the compact and poorly aerated soils. The penetration and development of root are poor. Such undeveloped root system cannot absorb sufficient moisture and nutrients from the soil.

Microorganism population and activity

The microorganisms living in the soil also require oxygen for respiration and metabolism. Some of the important microbial activities such as the decomposition of organic matter, nitrification, sulphur oxidation etc., depend upon the amount of oxygen present in the soil air. The deficiency of oxygen in soil air slows down the rate of microbial activity. The decomposition of organic matter is retarded and nitrification arrested. The microorganism population is also drastically affected by poor aeration.

Formation of toxic material

Poor aeration results in the development of toxins and other injurious substances such as ferrous oxide, H_2S gas, CO_2 gas, ethylene, organic acids, etc., in the soil.

Water and nutrient absorption

A deficiency of oxygen has been found to check the nutrient and water absorption by the plants. The energy of respiration is utilized in absorption of water and nutrients. Under poor aeration in water-logged soils, plants exhibit water and nutrient deficiency.

Development of plant diseases

Insufficient aeration of soil leads to the development of some diseases like, wilt of gram and dieback of citrus and peach.

Normal Aeration: In soil the normal rate of production of CO_2 is about 7litre/ m^2/day. With this rate of CO_2 production, the soil air would have to be completely removed to a depth of 20 cm every hour so that the O_2 content of the soil air remains at normal observed rate(20.65%). The replacement of soil air to a depth of 20 cm every hour is termed as Normal Aeration which is considered as a convenient standard to compare the relative aeration efficiency.

Measurement of soil aeration

Oxygen Diffusion Rate (ODR) is an electrical measurement that measures the electric current required for the reduction of all oxygen present at the surface of a cylindrical Pt-electrode in the soil. The flow of oxygen through the air-filled pores and the water film on the electrode is measured until the steady state is reached. ODR decreases with soil depth.

6

Measurement of Oxygen Diffusion Rate (ODR) and Redox Potential

Measurement of soil aeration

Oxygen Diffusion Rate (ODR) - the rate at which O_2 in soil air is replenished. ODR decreases with soil depth.

Measurement of ODR in soil

Lemon and Erickson (1952) devised a method for measuring oxygen diffusion in the soil system with the help of platinum micro electrode

Principle

When a certain potential is applied across the platinum electrode and a reference electrode inserted in the soil, oxygen is reduced at the platinum surface. The electric current flowing between two electrodes is proportional to the rate of oxygen reduction. The later, in turn, is related to the rate of oxygen diffusion to the electrode. The oxygen diffusion rate (ODR) is calculated from the measured electric current with the following equation.

$$ODR = Mi/\eta FA \qquad (6)$$

Where M is molecular wt of O_2, η is the number of electrons involved in the reduction of one molecule of oxygen, F is Faraday constant, i is current in amperes and A is the exposed surface (m^2) of the electrode.

Oxygen diffusion meter

The oxygen diffusion meter measures the mobility of oxygen in the soil that is important for the availability of oxygen for plants.

The method

Measuring the electric current required for the reduction of all oxygen present at the surface of a cylindrical Pt-electrode in the soil. The flow of oxygen through the air-filled pores and the water film on the electrode is measured until the steady state is reached.

The Oxygen Diffusion Rate (ODR) probe (Pt-electrode) should be placed in undisturbed soil. To this purpose a hole is predrilled to a depth of approximately 10 mm above the measuring point, after which the probe is lowered and carefully pushed into the bottom of the auger-hole. It is advised to remove the electrode from the soil after a series of measurements in order to clean it. The meter provides a stabilized voltage between the ODR-probe and the Ag-AgCl reference electrode.

In very dry soils only part of the electrode will be covered in water. This results in rising impedance between soil and electrode. In such a situation the meter can also be used to perform a redox-potential measurement.

The measuring system consists of a read-out unit with connecting facilities for three ODR-probes, one ODR-probe,(Pt electrode) one Ag-AgCl reference electrode, KCl-solution and a brass electrode. The reference electrode is used for measuring and checking the potential between the Pt-electrode and the soil. The brass electrode is used to close the electrical circuit. The measuring range for oxygen diffusion is 0 - 999 μA and for Redox 0 - 999 mV (resolution resp. 1 μA and 1 mV). Accuracy +/- 3 μA and +/- 3 mV. Operating temperature between 0 and 50°C and an air humidity between 30-80%.

Oxygen diffusion rate calculation

The instrument measures the electric current needed for a quantitative reduction of the oxygen flowing from the airspaces in the soil to the platinum electrode. The chemical reaction is given by:

$$O_2 + 2H_2O + 4e◊ = 4\ OH^- \qquad (7)$$

The reduction of one mole of oxygen thus requires four mole electrons which is equivalent to a charge of 4 x 96500 = 386000 Coulombs.

If ODR is the reduction rate of oxygen expressed in grams per second per square meter, this can be done as follows:

$$ODR = \frac{i \times 32}{386000 * A} \qquad (8)$$

In which i is the current readout in micro-amperes and A the surface area of the electrode in square meters. In the equation i should be used as a dimensionless value. Given the surface of the electrode is about $2.37*10^{-5}$ m^2(or 0.0000237 m^2) this results in:

ODR = i * 3.5 micrograms m^{-2} s^{-1}

in which **i** is the dimensionless value of the instrument readout.

So if you read the value 40 in the display ODR becomes 40* 3.5 = 140 microgramsm^{-2} s^{-1}

Implication of ODR towards crop growth

ODR <40x10^{-4} microgramsm^{-2} s^{-1} growth of plants appear to suffer.

Redox potential measurements

Redox potential is an electrical measurement that shows the tendency of a soil solution to transfer electrons to or from a reference electrode. From this measurement we can estimate whether the soil is aerobic, anaerobic,and whether chemical compounds such as Fe oxides or nitrate have been

chemically reduced or are present in their oxidized forms. Making these measurements requires three basic pieces of equipment: 1. Platinum electrode, 2. Voltmeter and 3. Reference electrode

The Pt wire is buried into the soil to be in contact with the soil solution. The reference electrode must also be in contact with the soil solution. It has a ceramic tip, which can be placed in the soil, or the electrode can be placed in a salt bridge, which is itself placed in the soil. Wires from both the Pt electrode and reference electrode are connected to the voltmeter.

Correction factors

While both kinds of reference electrodes give reliable data, the voltages measured with each electrode are interpreted slightly differently. It is for this reason that users must know which electrode they have.The voltage measured in the field must be corrected to what would have been obtained with a different reference electrode, called the standard hydrogen reference electrode. This electrode cannot be used in the field, but our interpretations of redox potential measurements are based on values determined with it. Therefore, all voltages measured in the field with either the Ag/AgCl or calomel reference electrode have to be adjusted to the value that would have been obtained a standard hydrogen electrode had been used.

The basic correction factors are:

Reference Electrode	Correction Factor (mV)
Ag/AgCl in saturated KCl solution	+200
Calomel	+250

These correction factors are temperature dependent, but in most instances the effect of temperature is must less than the variability in the data for a given time.

Anaerobic					Aerobic	Sediment condition
Highly reduced	Reduced		Moderately reduced		Oxidized	Redox condition
Co_2	$SO4^{2-}$	Fc^{3+}	Mn^4+	NO_3^-	O_2	Electron acceptor
Anaerobic	Facultative				Aerobic	Microbial metabolism

–300 –200 –100 0 +100 +200 +300 +400 +5C0 +600 +700

Redox potential range in soil and sediment showing the microbial metabolism process and electron acceptor.

Figure

An interpretation from figure

The E_h scale in mV as shown above is used to quantify the reduction or oxidation intensity in chemical or biological systems. Soil redox potential represents an indication of the oxidation reduction status of the various redox couples. For example, a redox potential of 0 mV indicates that the O_2 and NO_3 are not likely to be present and that bioreducible Fe and Mn compounds are in a reduced state. At the same potential however, SO_4 is stable in the soil with no production of sulphide, which is toxic to plants which is toxic to plants. A redox potential of + 400 mV indicates that O_2 may be present even though there may be excess water.

In ODR current is measured andused where as in Redox potential voltage developed is used.

References

Bornstein, J., W.E. Hedstrom, and F.R. Scott. 1980. Technical Bulletin on Oxygen Diffusion Rate Relationships in Three Soil Conditions. Life Sciences and Agriculture Experiment Station, University of Marine at Orono.

Daniel Hillel. Fundamentals of soil Physics. Academic Press.

Gildayal, B.P. and R.P. Tripathi. 1987. Soil Physics. Published by John Wiley & Sons (Asia) Pte Ltd., Singapore

Oswal, M.C. 1983. Soil Physics. Vikas Publishing House, New Delhi
Rattan Lal, R. Lal, Manoj Shukla. 2004. Principles of Soil Physics. CRC press
T.D. Biswas and S.K. Mukherjee.1995. Text book of Soil Science. Tata McGraw-Hill Publishing Company Limited, New Delhi
WWW. Optisci. Com

7

Use of Radio Tracer Techniques in Soil Fertility Evaluation & Radio Tracer Technology Application in Plant Nutrient Studies

In nature there are nearly 300 nuclei, consisting of different elements and their isotopes. Isotopes are nuclei having the same number of protons but different number of neutrons. These techniques are now being used in almost all the areas of plant physiology, soil chemistry and plant biochemistry. In agriculture, radioisotopes are used in the nutritional studies of major and minor elements, plant metabolism, mechanism of photosynthesis studies, and uptake of nutrients and ions mobility in soil. In applying this technique, a minute quantity of radioisotope element is usually mixed with ordinary element of the same kind, the whole batch is termed "tagged" and their role in chemical reactions is traced. The radio labeled fertilizer has been used to study the uptake, retention and utilization of fertilizers.

Radioisotopes are used for determining the function of fertilizer in different plants. The tracer technique is used to study the rate and direction of movement of an element in a plant. For this a radioisotope of that element is injected in the ground near the plant. After a few days the plant is laid on a photographic paper to produce an autoradiograph. The dark areas in the radiograph show the positions reached by the element. This technique gives valuable information regarding the optimum season for fertilizing crops and for poisoning weeds. Use of N^{15} in organic fertilizer studies has significantly advanced the understanding of N release from organic materials. Crop residue or green manure studies are relatively simple. These techniques have been extremely useful in determining how residue quality parameters affect mineralization values. Because of the attention focused on nutrient cycling and management studies in recent years, a number of new P isotope techniques have been developed.

Radioisotopes

The isotopes having unstable nuclei are known as radioisotopes. Their radioactive atoms are unstable because certain combinations of neutrons and protons produced nuclei of latent instability. Specific average life time will be characteristic of each unstable combination. The disintegration of a given radioactive atom is a matter of chance conditioned only by a set of many requirements being completely fulfilled simultaneously. They disintegrate spontaneously at a characteristic decay rate.

Type of radioisotopes

(i). *Stable Isotope:* No disintegration and no radioactivity

(ii). *Unstable Isotope:* Radioactive disintegration with emitted alpha, beta and gamma radiation. These unstable nuclei have excessive energy and this is released by their disintegration into stable forms of lower energy, the excess being released in the forms of radiations, mainly alpha (α), beta (β) and gamma (¥) and they are called radioactive.

Stable Isotopes in agriculture

Stable isotopes are used in the same way as radioactive isotopes in soil/plant studies. Whereas radioactive isotopes emit particles which are captured in photomultiplier tubes and counted, stable isotopes are separated from each other by passing a gas containing them through a strong magnetic field, which deflects them differentially according to their mass. The most common stable isotope used is N^{15}, but a large number of other stable isotopes are produced which are increasingly being used in agricultural studies.

Nitrogen is one of the main limiting factors for plant growth. There are twelve isotopes of nitrogen, many with extremely short half lives. Of the radioactive isotopes only N^{13} with a half-life of 9.97 minutes has been used mainly in plant nitrogen translocation experiments. Two stable isotopes of N occur naturally in the atmospheric N_2.

Use of radioisotope techniques in soil fertility evaluation

Fundamental assumption

"When a plant is confronted with two or more sources of a nutrient element, the nutrient uptake from each of these sources is proportional to the amounts available in each source". Implicit in this assumption is that there is complete mixing of the isotope or isotopic equilibrium. The above relationships, which are also called fractional utilization ratios, or isotopic dilution equation can be expressed in the form of an equation. In a situation where soil and fertiliser are the only sources of nutrient available to a plant, the equation is as follows:

$$\frac{Total\ nutrient_{plant}}{Total\ nutrient_{supply}} = \frac{Fertilizer\ nutrient_{plant}}{Fertilizer\ nutrient_{supply}} = \frac{Soil\ nutrient_{plant}}{Soil\ nutrient_{supply}}$$

Radioisotope techniques

(A) Tracer technique

The radio active isotopes can be measured in extremely minute quantities- one in one million- by sensitive electronic instruments and can also be recorded on a photographic plate and can be traced, hence as tracer techniques are the most indispensable tool in refined research in a wide range of nutritional, metabolic, developmental and pathological investigations in plants animals and man. Today, some important radioisotopes, such as P^{32}, C^{14}, Ca^{45}, H^{3}, Na^{22}, Co^{60}, Cu^{67}, Fe^{59}, Mn^{56}, K^{42}, Rb^{86}, S^{35} and Zn^{65} are available in sufficient quantities and have been widely used to investigate a number of problems in agriculture and other fields.

Using conventional chemical analysis techniques only ng-µg of a substance can be detected, but in tracer technology, one detected disintegration corrasponds to the decay of a single atom. Routine radioanalytical technique will allow detection of quantities that are 10^{5} times smaller than those needed for chemical analysis.Hence these are very accurate, reliable and satisfactory and solve the mysteries otherwise unexplorable. Applications of radioisotopes and radiations are helping us to find the solution of problems in much shorter time. These are economic, non destructive. True dynamics of various biochemical reaction can be studied. These are advantageous as radio tracing eleminates the artifacts of the chemical methods of estimation. Hence these are very accurate, reliable and satisfactory and solve the mysteries otherwise unexplorable. Applications of radioisotopes and radiations are helping us to find the solution of problems in much shorter time.

Tracer technique is based on the assumption that if a plant derives a particular nutrient from the soil as well as fertilizer added to it, the amount available from soil in terms of a standard fertilizer can be calculated if that derived from the fertilizers is known. The latter is possible by using with the fertilizer a radioactive or stable isotope of the nutrient in the question. Some plants are grown in soil to which is an added phosphatic fertilizer mixed with a very small amount of phosphate having radioactive P^{32}. The soil originally contains a known amount of non radioactive phosphorus, so does the bulk of phosphatic fertilizers, excepting the small amount of P^{32} added deliberately. After a suitable period of growth, the plants are harvested and the total P as well as P^{32} contents are determined using fractional utilization ratios.

Estimation of available nutrients in soil

Assessment of the available soil nutrient is of considerable importance in planning optimum fertilizer use. Out of several methods available to estimate the available soil nutrient, isotopic methods have received attention in this field.

L-value: It was first suggested by Larsen, 1952 that if labeled phosphate fertilizers was added to a soil at different rates and plants were grown, the specific activity analysis of the plant material give a constant value, if the isotope dilution of added chemical was assumed to take place in the soil system. The reason for the constant value obtained is attributed to the equilibrium which will be reached between added phosphate and the exchangeable phosphate in soil. The quantity measured in this way is known as L value and this quantity also termed as "Labile" phosphorus of the labile pool.

E-value: This method is directly application of the isotope dilution principle. The amount of nutrient in the soil at equilibrium with the same nutrient in the soil solution can be measured with E- value. Reaction can be depicted as fallows:

Surface P^{31} + Solution P^{32} Surface P^{32} + Solution P^{31}

At equilibrium,

$$\frac{\text{Surface } P^{31}}{\text{Surface } P^{32}} = \frac{\text{Solution } P^{31}}{\text{Solution } P^{32}}$$

In this method, it is possible to measure both capacity (in the adsorbed solid phase) and intensity (in the solution phase) factors of the same sample. The E-value represents labile form of nutrient and represents the total amount of nutrient undergoing isotopic dilution. The L- value can be considered as E-value in which the plant is being used as a means of sampling the solution upon equilibrium of added P with exchangeable soil P. The added advantage in this procedure is that the conditions are identical to those prevailing in the soil-plant system.

A- Value: The available amount of soil nutrient, or the A value of the soil for the particular nutrient under study, is expressed in equivalent units of the applied fertiliser. A- Value concept was developed by Fried, 1952. When the plant is confronted with two source of a given nutrient, the plant absorbs from each of these in proportion to the respective amount available. The amount of available nutrient in soil to be determined in terms of fertilizer standard is known as A-value. Mathematically A-value can be expressed as:

$$A = BX \frac{1-Y}{Y}$$

Where A = Amount of available nutrient in soil

B = Amount of applied fertilizer nutrient

Y = proportion of nutrient in the plant derived from the fertilizer nutrient.

B. Stable isotope Techniques

a. Uses of N^{15}labeling techniques

The two stable isotopes of nitrogen (N^{15} and N^{14}) naturally exist and their ratio in the atmosphere is constant(0.3663%), whereas it varies in the soil from 0.360% to 0.373%. This crop is then harvested , The % age of N^{15} ae(atom excess) is measured with a mass spectrophotometer coupled with elemental analyzer.

Organic Residues studies

(i) N^{15} direct labeling techniques:

Crop residue or green manure studies using the direct method are relatively simple. Green manures can be easily obtained by growing crops fertilized with N^{15} tracer; the aboveground or belowground material is then harvested and added as residue to unlabelled soil where the next crop is grown and the percentage nitrogen in the crop derived from added residue is calculated using the following Equation (Hauck and Bremner, 1976):

Atom % N^{15} excess in the crop

Ndfr (%) = X100

Atom % N^{15} excess in the added residue

Where Ndfr is nitrogen derived from fertilizer.

(ii) Indirect techniques:

1. Indirect techniques have been used to study plant N uptake from organic residues.

N^{15} tracer is added to the soil and treatments with and without residues (no-residue controls) are set up. The no residue controls will have a N^{15}enrichment that reflects the soil N^{15} pool and the residue treatments should have lower N^{15} enrichment due to the input of the unlabelled N coming from the decomposing residue. Nitrogen derived from residue (Ndfr) is calculated using the following

Equation.

Atom % N^{15} excess $_{\text{residue treatment}}$

Ndfr(%) = 100 X100

Atom % N^{15} excess $_{\text{no residue control}}$

In practice it has been shown that if N^{15} label and residues are applied at the same time this causes errors associated with pool substitution . Thus it is recommended that the soil be pre-labeled with N^{15} and left to equilibrate for up to 6 months prior to the application of residues.

(2) Biological N fixation

The same principle as the N^{15} dilution method for estimating N uptake from organic residues. The percentage of nitrogen introduced in the system **Nd atm**) is calculated by the isotopic dilution equation.

Atom % N^{15} excess $_{\text{treatment}}$

Nd atm (%) = 100 X100

Atom % N^{15} excess $_{\text{control}}$

b. Carbon-14 dating

Carbon dating is a method of obtaining age estimates on organic materials. This is often used for dating buried soils. Although a few measurements have been made on the contemporary soils. Radiocarbon dating estimates can be obtained on wood, charcoal, marine and freshwater shells, bone and antler, and peat and organic-bearing sediments.

Carbon (C) has three naturally occurring isotopes. Both C^{12} and C^{13} are stable, but C^{14} decays by very weak beta decay to nitrogen-14 with a half-life of approximately 5,730 years. Naturally occurring radiocarbon is produced as a secondary effect of cosmic-ray bombardment of the upper atmosphere. Plants transpire to take in atmospheric carbon, which is the beginning of absorption of carbon into the food chain. Movement of matter is traced to elucidate the reactions in photosynthesis. Animals eat the plants and this action introduces carbon into their bodies. After the organism dies, carbon-14 continues to decay without being replaced. To measure the amount of radiocarbon left in an artifact, scientists burn a small piece to convert it into carbon dioxide gas. Radiation counters are used to detect the electrons given off by decaying C^{14} as it turns into nitrogen. The amount of C^{14} is compared to the amount of C^{12}, the stable form of carbon, to determine how much radiocarbon has decayed, thereby dating the artifact.

The turnover of soil organic matter is usually too rapid for age measurement by using C^{14} dating technique particularly in the soils where the turnover is not restricted by desiccation, water logging and low temperature. S^{35} has proven to be an extremely useful tool in such studies because it has a convenient half-life (87.2 days).

Radioisotopes are used to improve the quality and productivity of agricultural products as well as optimum utilization of fertilizers without harmful effect to plants and mankind. The radio labeled fertilizer has been used to study the uptake, retention and utilization of fertilizers. Radioisotopes have played an important role in improving productivity in agriculture in a sustainable manner.

Further reading

Zapata F. Isotope Techniques in Soil Fertility and Plant Nutrition Studies

8

Soil Microorganisms and Their Importance

Microorganisms being minute and microscopic are ubiquitous. Besides supporting the growth of various biological systems, soil and soil microbes serve as a best medium for plant growth. Soil fauna & flora convert complex organic nutrients into simpler inorganic forms which are readily absorbed by plants.

Soil microbes as nitogen fixers-Symbiotic (Associative): The organisms involved are *Rhizobium, Bradyrhizobium* in legumes(aerobic): *Azospirillum, Azotobacter* (grasses), Actinonycetes- *Frantkia* (with Casuarinas, Alder).

Soil microbes as biocontrol agents: Several ecofriendly bioformulations of microbial origin are used in agriculture for the effective management of plant diseases, insect pests, weeds etc. eg: *Trichoderma* sp and *Gleocladium* sp are used for biological control of seed and soil borne diseases. Fungal genera *Entomophthora, Beauveria, Metarrhizium* and *protozoa- Maltesia grandis. Malameba locustiae* etc.- are used in the management of insect pests. Nuclear polyhydrosis virus (NPV) is used for the control of Heliothis/American boll worm. Bacteria like *Bacillus thuringiensis*, *Pseudomonas* are used in cotton against Angular leaf spot and bollworms.

Degradation of pesticides in soil: Soil receives different toxic chemicals in various forms and causes adverse effects on beneficial soil micro flora / micro fauna, plants, animals and human beings. Various microbes present in soil act as the scavengers of these harmful chemicals in soil. The pesticides/chemicals reaching the soil are acted upon by several physical, chemical and biological forces exerted by microbes in the soil and they are degraded into non-toxic substances and thereby minimize the damage caused by the pesticides to the ecosystem. For example, bacterial genera like *Pseudomonas, Clostridium, Bacillus, Thiobacillus, Achromobacter etc.* and fungal genera like

Trichoderma, Penicillium, Aspergillus, Rhizopus, and *Fusarium* play important role in the degradation of the toxic chemicals / pesticides in soil.

Biodegradation of hydrocarbons: Natural hydrocarbons in soil like waxes, paraffins, oils etc are degraded by fungi, bacteria and actinomycetes. *E.g.* ethane (C2 H6) is metabolized and degraded by *Mycobacteria*, *Nocardia, Streptomyces Pseudomonas*, *Flavobacterium* and several fungi.

Humus formation

During the course of their activities, the microorganisms synthesize number of compounds which play important role in humus formation.

Mineral availability

In addtion to their role in N and P availability (which will be discussed in detail later) they play avital role in the availability of other nutrients. Efficient potassium solubilising micro-organism strains including bacteria (*Bacillus mucilaginous*, *Bacillus edaphicus*, *Bacillus circulans*, *Acidothiobacillus ferroxidans*, *Paenibacillus* spp.) and fungi (*Aspergillus ferrous*) solubilse fixed forms of potassium (muscovite, biotite, orthoclase and feldspar) through acidolysis, chelation. Exchange reactions, complexolysis and production of organic to their soluble form.

Iron (Fe) is one of the most limiting nutrients due to the low solubility of oxidised ferric form in aerobic environments. Common strains found to increase Fe availability are *Trichiderma asperellum*, *Methylobacterium oryzae*, *Pseudomonas putida* and *Enterobacteria* spp. Secretion of low molecular organic compound (siderophores) by *Pseudomonas* carry Fe to the root surface.

Zinc (ZN) solubilization is carried out by numerous bacteria example *Acinetobacter* spp. and *Bacillus aryabhattai*. The microorganisms solubilise Zn by reducing rhizospheric pH or sequester the cations to which it is bound to increase Zn availability.

Gluconic acid is considered to be the major organic acid involved in the solubilization of insoluble minerals.

The role of various groups of microbes in soil is concisely described below.

(A) Bacteria

Among the different microorganisms inhabiting in the soil, bacteria are the most abundant and predominant organisms. These are primitive, prokaryotic, microscopic and unicellular microorganisms without chlorophyll. Bacteria bring about a number of changes and biochemical transformations in the soil and thereby directly or indirectly help in the nutrition of higher plants growing in the soil. The important transformations and processes in which soil bacteria play vital role are: decomposition of cellulose and other carbohydrates, ammonification (proteins -ammonia), nitrification(ammonia-nitrites-nitrates), denitrification (release of free elemental nitrogen), biological fixation of atmospheric nitrogen (symbiotic and non-symbiotic), oxidation and reduction of sulphur and iron compounds. All these processes play a significant role in plant nutrition,

Process/reaction	Bacterial genera
Cellulose decomposition (celluloytic bacteria) most cellulose decomposers are mesophilic	a. Aerobic : *Angiococcus, Cytophaga, Polyangium, Sporocytophyga, Bacillus, chromobacter, Cellulomonas* b. anaerobic: *Clostridium Methanosarcina, Ethanococcus*
Ammonification (Ammonifiers)	*Bacillus, Pseudomonas*
Nitrification (Nitrifying bacteria)	*Nitrosomonas, Nilrobacter ,Nitrosococcus*
Denitrification (Denitrifiers)	*Achromobacter, Pseudomonas, Bacillus Micrococcus*
Nitrogen fixing bacteria	a) Symbiotic- *Rhizobium, Bradyrrhizobium* b) Non-symbiotic: aerobic *Azotobacter Beijerinckia* (acidic soils), anaerobic- *Clostridium*

Various plant residues in soil are degraded by bacteria *viz.*

Cellulose (*Pseudomonas,Cytophaga, Spirillum, Cellulomonas,* Actinomycetes), Hemicelluloses (*Bacillus, Vibrio, Pseudomonas, Erwinia)* Lignin (*Pseudomonas, Micrococcus, Flavobacteriumm, Xanthomonas, Streptomyces)* Pectin (*Erwinia*) and Proteins (*Clostridium, Proteus, Pseudomonas, Bacillus*) .

(B) Fungi

Fungi in soil are present as myceliul bits, rhizomorph or as different spores. Their number varies from a few thousand to a few -million per gram of soil. Soil fungi possess filamentous mycelium composed of individual hyphae. Most commonly encountered genera of fungi in soil are; *Alternaria, Aspergillus, Cladosporium, Cephalosporium Botrytis, Chaetomium, Fusarium, Mucor, Penicillium, Verticillium, Trichoderma, Rhizopus, Gliocladium, Monilia, Pythium, etc.* Most of these fungal genera belong to the subdivision Deuteromycotina / Fungi imperfeactai which lacks sexual mode of reproduction. As these soil fungi are aerobic and heterotrophic, they require abundant supply of oxygen and organic matter in soil. Fungi are dominant in acid soils, because acidic environment is not conducive / suitable for the existence of either bacteria or actinomycetes. The optimum PH range for fungi lies-between 4.5 to 6.5.

They are also present in neutral and alkaline soils and some can even tolerate PH beyond 9.0.

Role of Fungi

Fungi play significant role in soils and plant nutrition.

1. They play important role in the degradation / decomposition of cellulose, hemi cellulose, starch, pectin, lignin in the organic matter added to the soil.
2. Lignin which is resistant to decomposition by bacteria is mainly decomposed by fungi.
3. They also serve as food for bacteria.
4. Certain fungi belonging to sub-division Zygomycotina and Deuteromycotina are predaceous in nature and attack on protozoa & nematodes in soil and thus, maintain biological equilibrium in soil.
5. They also play important role in soil aggregation and in the formation of humus.
6. Some soil fungi are parasitic and cause number of plant diseases such as wilts, root rots, damping-off and seedling blights eg. *Pythium, Phyiophlhora, Fusarium, Verticillium* etc.
7. A number of soil fungi form mycorrhizal association with the roots of higher plants (symbioticassociation of a fungus with the roots of a higher plant) and help in mobilization of soil phosphorus and nitrogen eg. *Glomus, Gigaspora, Aculospora* (Endomycorrhiza) and *Amanita, Boletus, Entoloma, Lactarius* (Ectomycorrhiza).

(C) Actinomycetes

These are the organisms with characteristics common to both bacteria and fungi but yet possessing distinctive features to delimit them into a distinct category. In the strict taxonomic sense, actinomycetes are clubbed with bacteria the same class of Schizomycetes and confined to the order Actinomycetales. They are unicellular like bacteria, but produce a mycelium which is non-septate (coenocytic) and more slender, like true bacteria they do not have distinct cell-wall and their cell wall is without chitin and cellulose (commonly found in the cell wall of fungi). On culture media unlike slimy distinct colonies of true bacteria which grow quickly, actinomycetous colonies grow slowly, show powdery consistency and stick firmly to agar surface. They produce hyphae and conidia/ sporangia like fungi. Certain actinomycetes whose hyphae undergo segmentation resemble bacteria, both morphologically and physiologically. Actinomycetes are numerous and widely distributed in soil and are next to bacteria in abundance. They are widely distributed in the soil, compost *etc*. Plate count estimates give values ranging from 10^4 to 10^8 per gram of soil. They are sensitive to acidity/

low PH (optimum PH range 6.5 to 8.0) and waterlogged soil conditions. The population of actinomycetes increases with depth of soil even up to horizon C of a soil profile. They are heterotrophic, aerobic and mesophilic (25-30° C) organisms and some species are commonly present in compost and manures are thermophilic growing at 55-65° C temperature (*e.g.* Thermoatinomycetes, Streptomyces). Actinomycetes belonging to the order of Actinomycetales are grouped under four families *viz.* Mycobacteriaceae, Actinomycetaceae, Streptomycetaceae and Actinoplanaceae. Agriculturally and industrially important actinomycetous genera are present in only two families of Actinomycetaceae and Strepotmycetaceae. In the order of abundance in soils, the common genera of actinomycetes are Streptomyces (nearly 70%), *Nocardia* and *Micromonospora,* although Actinomycetaceae, Actinoplanes, *Micromonospora* and *Streptosporangium* are also generally encountered.

Role of actinomycetes

1. Degrade/decompose all sorts of organic substances like cellulose, polysaccharides, protein fats, organic-acids etc.
2. Organic residues/substances added to soil are first attacked by bacteria and fungi and later by actinomycetes, because they are slow in activity and growth than bacteria and fungi.
3. They decompose/degrade the more resistant and indecomposable organic substance/matter and produce a number of dark black to brown pigments which contribute to the dark colour of soil humus.
4. They are also responsible for subsequent further decomposition of humus (resistant material)in soil.
5. They are responsible for earthy/musty odor/smell of freshly ploughed soils.
6. Many genera species and strains (eg. Streptomyces if actinomycetes produce/ synthesize a number of antibiotics like Streptomycin, Terramycin, Aureomycin etc.
7. One of the species of actinomycetes *Streptomyces scabies* causes disease "Potato scab" in potato.

(D) Algae

Algae are present in most of the soils where moisture and sunlight are available. Their number in soil usually ranges from 100 to 10,000 per gram of soil. They are photoautotrophic, aerobic organisms and obtain CO_2 from atmosphere and energy from sunlight and synthesize their own food. They are unicellular, filamentous or colonial. Soil algae are divided in to four main classes or phyla- 1. Cyanophyta (Blue-green algae), 2. Chlorophyta (Grass-green algae), 3. Xanthophyta (Yellow-green algae) and 4. Bacillariophyta (diatoms or golden-brown algae). Out of these four classes/phyla, blue-green algae and grass-green algae are more abundant in soil. The green-grass algae and diatoms are dominant in the soils of temperate region while blue-green algae predominate in tropical soils. Green-algae prefer acid soils while blue green algae are commonly found in neutral and alkaline soils. The most common genera of green algae found in soil are: *Chlorella, Chlamydomonas, Chlorococcum, Protosiphon* etc. and that of diatoms are *Navicula, Pinnularia. Synedra, Frangilaria.* Blue green algae are unicellular, photoautotrophic prokaryotes containing Phycocyanin pigment in addition to chlorophyll. They do not posses flagella and do not reproduce sexually. They are common in neutral to alkaline soils. The dominant genera of BGA in soil are: *Chrococcus, Phormidium, Anabaena, Aphanocapra, Oscillatoria* etc. Some BGA posses specialized cells know as "Heterocyst" which is the sites of nitrogen fixation. BGA fixes nitrogen (non-symbiotically) in puddle paddy/water logged paddy fields (20-30 kg/ha/season). There are certain BGA which possess the character of symbiotic nitrogen fixation in association with other organisms like fungi, mosses, liverworts and aquatic ferns Azolla, eg *Anabaena-Azolla* association fix nitrogen symbiotically in rice fields.

Role of algae or BGA

1. Play important role in the maintenance of soil fertility especially in tropical soils.
2. Add organic matter to soil when die and thus increase the amount of organic carbon in soil.
3. Most of soil algae (especially BGA) act as cementing agent in binding soil particles and thereby reduce/prevent soil erosion.
4. Mucilage secreted by the BGA is hygroscopic in nature and thus helps in increasing water retention capacity of soil for longer time/period.
5. Soil algae through the process of photosynthesis liberate large quantity of oxygen in the soil environment and thus facilitate the aeration in submerged soils or oxygenate the soil environment.
6. They help in checking the loss of nitrates through leaching and drainage especially in un-cropped soils.
7. They help in weathering of rocks and building up of soil structure.

(E) Virus

They cause disease in other microbes and plants.

9

Saline, Alkali, Acid, Waterlogged and Sandy Soils: Their Appraisal and Management

There are four major tracts where salt affected soils are commonly met within India. These are:

1. The Semi-arid Indo-Gangetic alluvial tracts (mainly in Punjab, Haryana, Uttar Pradesh and a part of Bihar)
2. The arid tracts of Rajasthan and Gujarat.
3. The arid and semi arid tracts of southern states, particularly of the irrigated rigor (*Vertisols*) soils.
4. The coastal alluvium

Earlier it was estimated that about 7 million hectares of land have been affected by salinity/sodicity conditions in India. The area under these soils increased and reached to the level 10 million hectares at present.

Classification of saline sodic soils

Baed on pH, electrical conductivity (EC) and Exchangeable sodium percentage

(ESP = Exchangeable Na+ x 100/ CEC) saline –sodic soils are classified as follows:

Soil	EC(dsm-1)	ESP	pH
Saline	> 4.0	< 15	< 8.5
Sodic(Alkali)	< 4.0	> 15	> 8.5
Saline – sodic(Alkali)	> 4.0	>15	> 8.5

Saline soils / white alkali / solonchak

Soils with high amount of soluble salts having EC > 4.0 dsm^{-1} and white encrustations are seen on the surface. Hence it is called as white alkali. Saline soils with high proportion of nitrate salts are called brown alkali .

Genesis of saline – sodic soils

Parent material

Soils formed from rocks having high proportion of bases are become saline / sodic in nature. eg. Basalt, Sand stone etc.

Low rainfall

One of the important reason for the development of saline-sodic soils is insufficient water to remove bases from soil horizon and thereby accumulation of salts in soil. This is more common in semi arid and arid regions where the rainfall is usually low.

High evaporation

Water along with salts reaches the surface from sub surface of the soil by capillary raise due to high evaporation in arid and semi arid regions. This results in accumulation of salt at surface of the soil while water alone moves to atmosphere.

Poor drainage

Water logged salinity / sodicity is a common seen in low-lying area of inlands particularly in high clay soils. Improper drainage leads to accumulation of salts at surface horizon and becomes reason for entry of sodium in clay complex.

Poor quality irrigation waters

Continuous use of poor quality saline / sodic water for cultivation accumulates salts / sodium in the soils.

High water table

High water table at alluvial plains and other areas leads to improper drainage, which leads to accumulation of salts in soils.

Sea water intrusion

In coastal regions seawater intrudes into land and pollutes the soil as well as ground water of that locality.

Base forming fertilizers

Continuous application of base forming fertilizers for cultivation is also causes soil salinity / sodicity. eg. $NaNO_3$

Saline Soils

Soils having higher proportion of soluble salts affect adversely the growth of plants. The salt level in saline soils exceeds a limit of 4.0 dSm-1. Mostly these soils are dominant with chlorides and sulphates salts. These salts are neutral salts and hence the pH of these soils may not be more than 8.5. Saline soils are formed through a soil forming process called salinization in semi arid and arid zones. Salinization refers to accumulation of soluble salts in the soil surface horizons.

Effects of soil salinity

The characteristics feature of saline soil is white encrustation on surface of soils due to evaporation of water to atmosphere leaving the salts on surface of soils. Presence of salts leads to alteration of osmotic potential of the soil solution. Consequently water intake by plants restricted and there by nutrients uptake by plants are also reduced. In these soils due to high salt levels microbial activity is reduced consequently slow decomposition and low nutrient availability particularly nitrogen and sulphur. Due to osmotic potential alteration water from plants cells moves to soil and plants are affected due to dehydration. As a result drying of leaves and finally death of plants occur. Apart from above effects specific ion effects on plants are also seen due to toxicity of ions like chloride, sulphate etc.

Reclamation

All saline soils can be reclaimed easily of good quality water is available. Since the salts in this soils are soluble in nature using quality water they can be solubilized and leached off from the field. In the absence of good quality water in becomes necessary to manage saline soils for better growth of plants.

Management of saline soils

1.Crop management

Growing crops that are tolerant high level soil salinity e.g.: cotton, ragi, barley, sugar beat, beet root, curry leaf, bermuda grass, saline grass, spinach *etc*. Crops that are tolerant to soil salinity at medium level are paddy, wheat, onion, maize, sunflower, castor, grape, pomegranate, tomato, cabbage and potato. Crops that are tolerant to low level of soil salinity are garden beans, reddish, lime *etc*.

Black gram, green grams are sensitive to soil salinity. Crops are to be chosen based on the soil salinity level.

Relative tolerance of crops to salinity

Plant species	Threshold salinity (dS m-1)
Field crops	
Cotton	7.7
Sugar beet	7.0
Sorghum	6.8
Wheat	6.0
Soybean	5.0
Groundnut	3.2
Rice	3.0
Maize	1.7
Sugarcane	1.7
Vegetables	
Tomato	2.5
Cabbage	1.8
Potato	1.7
Onion	1.2
Carrot	1.0
Fruits	
Citrus	1.7

2.Soil / cultural management

Growing crops in raised beds will reduce accumulation of salt around root zone. Planting seedlings / sowing seeds on sloppy ridges decreases accumulation of salts around root zone. Mulching soil prevents evaporation which reduces accumulation of salts due to capillary rise of water at surface of soils. Providing drainage in water logged areas also helps to reduce salt accumulation.

3.Fertilizer management

Addition of extra dose of nitrogen to the tune of 20 – 25% of recommended level will compensate the low availability of N in these soils. Addition of organic manures like, FYM, compost, etc helps in reducing the ill effect of salinity due to release of organic acids produced during decomposition. Green manuring (Sunhemp, Daincha, Kolingi) and / or green leaf manuring also counteracts the effects of salinity.

4. Irrigation management

Proportional mixing of good quality (if available) water with saline water and then using for irrigation reduces effect of salinity. Alternate furrow irrigation favors growth of plant than flooding. Drip and sparinkler irrigation systems aim to reduce the use of water which is favorable for growth of plant since slat accumulation also reduced with low usage of water.

All the above four management practices suitably integrated to reduce the soil salinity, which is favorable for better growth of plants and ultimately for better yields. Management of saline soils becomes essentials and unavoidable particularly in areas where both soil as well as irrigation water are saline in nature.

Sodic soils

Sodic soils are having high proportion of sodium at exchange complex. The sodium ion at exchange complex usually exceeds 15 percentage in these soils. These soils also have high proportion of precipitated insoluble carbonates and bicarbonates and hence the pH always more than 8.5. On contrary degraded sodic soils have low pH at surface but exchangeable sodium percentage is more than 15 and they do not have precipitated $CaCO_3$. Sodic soils are formed due to the soil forming process of alkalization (accumulation sodium in soils) while solodi solids / degraded alkali (sodic) formed by the process called solidization.

Effect of soil sodicity

Since these soils have high amount of CO_3 and HCO_3 and high pH, Carbonate, bicarbonate and OH (hydroxyl) ions injuries on plants are observed. High sodium in clay becomes reason for dispersed nature of clay under wet moisture regions.

Dispersed nature of the clay leads to soapy feeling of soils, stagnation of water, poor infiltration/ percolation and poor aeration. Sodic soils become hard mass during dry periods. These soils have poor workability both under wet and dry seasons.

Further hazardous effects of Na on plants are also seen. Sodium carbonate with water releases Na+, HCO_3^- and OH- ions, which are harmful to growing plants.

$$2\ Na^+ + CO_3^{2-} + H_2O \rightarrow 2\ Na^+ + HCO_3^- + OH^-$$

High pH is also unfavorable for the growth of microorganisms. Low microbial activity causes slow decomposition of organic matter and hence nutrient availability is also affected specifically nitrogen, sulphur etc. Since these soils enriched with high Na at exchange complex, Ca and Mg availability are also less. High pH becomes the reason for non-availability or less availability of Fe and Zn. Deficiency of Zn is common in this soils. hosphorus availability is also less due to conversion of phosphors into insoluble calcium and magnesium phosphates. All the above effects on plants result in drying of plants in patches in a field. Under extreme conditions no plants are seen in these soils.

Materials

Gypsum, Calcium Chloride, Calcium Carbonates, Phospho gypsum etc are used for reclamation which are directly supply calcium as they have calcium in their composition. Among them gypsum is most commonly used. $CaCO_3$ is insoluble in nature which of no use insodic soils (have already precipitated $CaCO_3$) but can be used in degraded sodic soils (do not have precipitated $CaCO_3$) since pH of this soils are low and favoring solubilization of $CaCO_3$.

Some of the indirect suppliers of Ca (elemental sulphur, sulphuric acid, iron sulphate) are also used for sodic soils. These materials on application solubilize precipitated $CaCO_3$ in sodic soilsand releases Ca for reclamation (for exchange reaction). Since degraded sodic soils do not have precipitated $CaCO_3$, use of these materials is not beneficial. Lime sulphur supply Ca both directly as well as indirectly.

Reactions: (Direct sources)

$^{Na}\text{Micelle}^{Na} + CaSO_4 Þ = \text{Micelle}^{Ca} + Na_2SO_4$ (leachable)

$^{Na}\text{Micelle}^{Na} + CaCl_2 Þ = \text{Micelle}^{Ca} + 2NaCl$(leachable)

Indirect sources

Sulphur and iron sulphate are converted into sulphuric acid which dissolves native $CaCO_3$. Thus Ca is made available for exchange reactions. Sulphur – microbial oxidation

$$2S + 3O_2 \rightarrow 2SO_3$$

$$SO_3 + H_2O \rightarrow H_2SO_4$$

$$H_2SO_4 + 2CaCO_3 \rightarrow CaSO_4 + Ca(HCO_3)_2$$

$$2Na - X + CaSO_4 \rightarrow Ca\text{-}X + Na_2SO_4$$

Relative tolerance of crops to sodicity

ESP (range*)	Crop
2-10	Deciduous fruits, nuts, citrus, avocado
10-15	Safflower, black gram, peas, lentil, pigeon pea

10

Chemical and Mineralogical Composition of Important Horticultural Crops

The living plant is constituted of various organic compounds, which in turn are made up of several inorganic elements. The organic compounds to which the plant can be resolved at the first disintegration are called proximate constituents. The inorganic elements to which the plant can be finally decomposed are called ultimate components.

Proximate constituents are as follows

1. Water	:	80-95%
2. **Carbohydrates**a. Sugars and starchesb. Hemicellulosesc. Celluloses	:	1-5 %10-30 % 20-30 %
3. Proteins	:	1-15 %
4. Lipids	:	1-8 %
5. Plant pigments, alkaloids, tannins and essential oils	:	1-8 %
6. Plant growth substances (enzymes, vitamins and hormones)	:	Small amounts
7. Mineral elements	:	2-5 %

Ultimate components are: the inorganic elements which make up the plant body *Viz*. C, H, O, N, P, K, Ca, Mg, S, Fe, Mn, Zn, Cu, B, Mo and Cl.

Chemistry of vegetable crops

Vegetables play an important role in human nutrition supplying some of the materials, which other food materials are lacking. Vegetables are important in neutralizing the acidity formed during digestion. They also serve as roughages in human nutrition and thus helping digestion.

They are the important sources of minerals like Ca, P and Fe. They also form an important source of vitamins, especially vitamin A. Green and yellow vegetables like carrot, turnip, spinach, beans, etc. contain appreciable amounts of vitamin A. Direct seed of beans, peas, and legumes contain appreciable amounts of proteins. Vegetables also contain sufficient amounts of vitamin C and also appreciable quantities of thiamine, niacin and folic acid. Tomatoes and potatoes contain fairly high amounts of vitamin C.

Potato (*Solanum tuberosum*)

Starch accounts for 65-80% of the dry weight of the tuber. The important sugars are glucose, fructose and sucrose. The N content varies from 1.2 to 2.0 per cent on dry weight basis. Of the total N, less than half is present as protein. 35-60% of N consists of free amino acids, oxides and nitrogenous bases. The *principal protein* of potato is a *globulin* fractionated into two fractions- *tuberin and globulin II*. The enzymes are phosphorylases, α and β amylases and phosphatases.

Potatoes are good source for vitamin A and C. Riboflavin, thiamine and nicotinic acid are also present. The average fat content of the potato tuber is 0.1% on fresh weight basis. Citric and malic acids are the important organic acids present in potato. Solanine is the term applied to *steroidal glycoalkaloid of potatoes*.

Tomato (*Solanum lycopersicum*)

It is popularly praised vegetable commonly known as 'Love apple'. The nutrient content / 100 g of tomato are as follows:

Water	94.1%		
Protein	1.0%		
Fibre	0.6%		
Fat	0.3%		
Carbohydrate	4.0%		
Minerals			
Na	3 mg	Fe	0.6 mg
K	268 mg	Cu	0.1 mg
Ca	11 mg	Mn	0.19 mg
Mg	11 mg	P	27 mg
S	11 mg	Cl	51 mg
Vitamins			
Vit. A	1100 IU		
Vit. C	23 mg		
Vit. E	0.27 mg		
Vit. B	0.2 mg		
Nicotinic acid	0.6 mg		
Sugar content is 1.85 – 4.27%.			
Acidity	4.2 - 10.2 mg of acid / 100 ml		
Ascorbic acid	20.9 - 22.5 mg / 100 g		

It is an excellent source of Vitamin C, so it is commonly known as poor man's orange. It is popular salad vegetable. It is used for the preparation of soups, pickles, ketchups *etc*. Tomato juice is a popular appetizer and beverage.

Carrot (*Daucus carota*)

The nutrient content of the carrot per 100 g edible portion is as follows:

Water	82.2%
Energy	45 calories
Protein	1.2%
Vit. A	12,000 I.U
Thiamine	0.042 mg
Riboflavin	0.043 mg
Niacin	0.21 mg
Ca	42 mg
Vit. C	4 mg

Beetroot (*Beta vulgaris*)

100 g of fresh Egyptian table beet rootcontains

Protein	1.46 g	CHO	9.03 g
Fat	0.09 g	Ca	27 mg
Fibre	0.08 g	Cl	56 mg
Ash	0.89 g	Vit. A	19 I.U
Fe	0.8 mg	Vit. B	0.021 mg
Vit. B1	0.049 mg	Vit. C	5.5 mg

The tops are rich in vitamin and high in Ca and Fe. Small beets are excellent green leaf vegetable.

Cowpea (*Vigna sinensis* L.)

The nutrients contents of cowpea are given per 100 g edible portion

Moisture	84.6%
Fat	0.2 g
Fibre	2.0 g
Calories	51
P	74 mg
Vit. A	941 IU
Riboflavin	0.09 mg
Vit. C	13 mg
Protein	4.3 g
Minerals	0.9 g
CHO	8.0 mg
Ca	80 mg
Fe	2.5 mg
Thiamine	0.07 mg
Nicotinic acid	0.9 mg

Field beans (*Dolichos lablab*)

Dolichos bean is rich in its nutritive value. The average chemical composition is a follows:

Moisture	86.1 g	Vit. A	312 I.U
Fat	0.7 g	Riboflavin	0.06mg
Calories	48	Ca	210 mg
Fibre	1.8 g	Oxalic acid	1 mg
Protein	3.8 g	Fe	1.7 mg
Minerals	0.9 g	K	74 mg
CHO	6.7 g	S	40 mg
Mg	34 mg	Thiamine	0.1 mg
P	68 mg	Nicotinic acid	0.7 mg
Na	55.4 mg	Vit. C	9 mg
Cu	0.13 mg		

It is primarily grown for green pods, which are cooked as vegetables like other beans. The dry bean seeds also collected for various vegetable purposes.

Onion (*Allium cepa*)

The chemical composition per 100g of onion is as follows

Moisture	86.8%
Protein	1.2%
Fat	0.1%
CHO	11.5%
Ca	0.18%
P	0.05%
Fe	0.7 mg
Vit. B	80 mg
Riboflavin	10 mg
Nicotinic acid	0.4 mg
Ascorbic acid	11.0 mg

Immature and mature bulbs are used as vegetable. The pungency, which is due to a volatile oil known as allylpropyl disulphide.

Cabbage (*Brassica oleracea* var. capitata)

The chemical composition per 100 g of fresh cabbage is detailed below

Water	92.1 g	Vit. B2	0.04 mg
Protein	1.4 g	Vit. B6	0.11 mg
Total fats	2 g	Vit. C	46 mg
Total CHO	5.7 g	P	28 mg
Fibre	1.5 g	Ca	46 mg
Vit. A	70 I. U	K	227 mg
Vit. B_1	0.04 g	Na	13 mg

Cauliflower (*Brassica oleracea* var. botrytis)

The composition per 100 g of cauliflower is as follows

Water	91.7%	Vit. A	40 I. U
Energy	31 calories	Ascorbic acid	70 mg
Protein	2.4 g	Thiamine	0.2 mg
Ca	22 mg	Riboflavin	0.1 mg
Niacin	0.57 mg		

Cauliflower seedlings are used for salad and green. The curd is used in curries, soups and pickles. In abundant areas of production cauliflower curd is cut into pieces, dried and procured for off-season use.

Radish (*Raphanus sativus*)

The nutritive value of radish roots per 100 g of edible portion is as follows

Moisture	94.4 g	Ca	50 mg
Protein	0.7 g	Oxalic acid	9 mg
Fat	0.1 g	P	22 mg
Minerals	0.6 g	Fe	0.4 mg
Fibre	0.8 g	Na	33 mg
CHO	3.48 g	K	138 mg
Calories	17	Vit. A	5 I.U
Thiamine	0.06 g	Riboflavin	0.02 mg
Nicotinic acid	0.5 mg	Vit. C	15 mg
In leaves			
Water	89.1%	P	0.06%
Protein	3.9%	Fe	0.8 mg
Fat	0.6%	Vit. A	81 I.U
Nicotinic acid	1.4 mg	Riboflavin	2.7 mg
CHO	4.1%	Vit. B	21 mg
Ca	0.31%	Vit. C	21 mg

Fleshy roots are eaten raw or in salad or cooked. Fruits cooked and used as vegetables. It is very tasty when both roots and leaves are cooked together.

Moringa (*Moringa oleifera*)

The composition of moringa is as follows

Moisture	75.0%	Iodine	51 mg / kg
Protein	6.7%	Carotene	11.300 I.U
Fat	1.7%	Vitamin B1	210 mg / g
CHO	13.4%	Nicotinic acid	0.8 mg / 100 g
Fibre	0.9%	Ascorbic acid	220 mg / 100 g

Contd.

Mineral matter	2.3%	Tocopherol	7.4 mg / 100 g
Ca	440 mg / 100 g	Fe	7.0 mg / 100 g
P	70 mg / 100 g		

Leaf protein contains the following amino acids: (mg / 100 g N)

Arginine	6.0	Methionine	2.0
Histine	2.1	Threonine	4.9
Lysine	4.3	Leucine	9.3
Tryptophan	1.9	Isoleucine	6.3
Phenylalanine	6.4	Valine	7.1

Kernel of seed contains

Moisture	4.0 %
Crude protein	38.4 %
Fatty oil	34.7 %
Nitrogen free extract	16.4 %
Fibre	3.5 %
Mineral matter	3.2 %

The fruits, leaves and flowers are used in culinary preparations. Immature fruits are cut into pieces and used in several culinary dishes after scraping of tough skin. The seed yields valuable oil known as 'oil of ben' valued by watchmakers for lubricating delicate machines.

Chemical composition of vegetables

S. No	Crop	Moisture (%)	Proteins (%)	CHO's	Fats (%)	Ca	P (mg/100g)	Fe
1.	Chillies	85.7	2.9	3.0	0.6	30	24	1.2
2.	Brinjal	92.7	1.1	5.5	0.2	0.15	0.48	0.01
3.	Bottle guard	96.4	0.2	2.5	0.1	20	5	0.7
4.	Pumpkin	92.6	1.4	4.6	0.1	10	14	0.7
5.	Bhendi	89.6	1.9	3.4	0.2	60	43	1.5
6.	Turnip	91.6	0.5	6.2	0.2	30	-	0.4
7.	Garlic	62.0	6.3	29.0	0.1	30	-	1.8
8.	Knoolkhol	92.7	1.1	3.8	0.2	20	18	0.4
9.	Bitter gourd	83.2	2.1	9.8	1.0	50	21	9.4
10.	Amaranthus	85.0	4.0	6.3	0.5	397	247	25.5

Chemistry of fruits

Fruits are valued for their attractive appearance, flavour and texture for a long time. In recent years their vitamin content has been recognized as an important feature. In all these characteristics, sugars either in the free stage or as derivatives play an important role. Flavour is fundamentally the result of the balance between sugars and acids. Certain flavouring constituents which are mostly glucosides also add to the flavour of the fruits.

I. Sugars

The sugar content varies from traces to 61% in fruits. Traces of sugars are present in lime and 61% in date. Sugars other than glucose, fructose and sucrose are rarely present. The average sugar content varies from 5-10%.

Apple	6-16%
Pineapple	8-18%
Grapes	10-19%
Mango	14%
Date	61%
Tomato	2-4%
Lemon	0.9-3.6

II. Proteins

The protein content of fruits is comparatively very low. The protein content varies with species, locality, season, cultural practices and other environmental factors.

Apple	0.2%	Grapes	1.3%
Banana	1.1%	Guava	0.8%
Avocado	2.1%	Lime, mango and pineapple	<1%
Tomato	1.2%	Dates	2.2%

III. Volatile compounds

The characteristic odour of many fruits is due to the presence of certain volatile compounds. These volatile compounds of the fruits are usually less than 100 ppm in concentration.

This group generally includes esters but compounds such as alcohols, aldehydes, ketons *etc.* are also found to possess pleasant aroma.

Apple	Ethyl 2 methyl butyrate
Grapes	Methyl anthranilate
Banana	Amyl acetate and isopentyl acetate
Grape fruit	Terpenes

Lemon: Hydrocarbon containing isoprene units and their oxygenated derivatives.

Orange: Limonene.

IV. Fruit phenolic compounds

Phenolic compounds are responsible for colour flavour and taste. These compounds give both desirable and undesirable qualities in fruits. The fruit phenolic compounds include both "Flavonoids" and "Cinnamic acid". The major flavonoids are anthocyanin, leucoanthocyanin, flavones and flavonols. Cinnamic acid and its derivatives are not flavonoids but related to them. The presence of these phenolic substances in fruits gives an astringemt taste.

V. Fruit pigments

The chief pigments of fruit which impart colour are the carotenoids, chlorophylls, anthocyanin, anthoxanthin etc. The yellow, orange and red colour of mango, papaya, peach, apricot, tomato, red pepper, carrot, etc. are mainly due to carotenoids. The carotenoids containing hydroxyl groups are called xanthophylls. This is the specific colour-bearing constituent of yellow maize, papaya and mandarin oranges.

VI. Vitamins

The most important vitamin of a majority of fruits is vitamin-C. It is a sugar derivative.

Guava	200 mg/100g	Banana	10-30 mg/100g
Melons	25-35 mg/100g	Oranges	50 mg/100g
Apple	2-10 mg/100g	Tomato	25 mg/100 g

Some fruits also contain vitamin-A. β-carotene, the precursor of vitamin A or provitamin A is also present in some fruits. Some fruits also have nicotininc acid, folic acid, thiamine, etc.

References

Bradbur, J.H. and Holloay, W.D. 1994. Chemistry of Tropical Root Crop- significance for nutrition and agriculture in the Pacific. ACIAR monograph No. 0,201 p.

Browne, Charles Albert, Waltham, Mass. 1944. A Source Book of Agricultural Chemistry. Published by The Chronica Botanica Co.; New York City, G.E. Stecher, 1944.

M. Dhakshinamoorthy, 2000. An Introduction to Plant Bio Chemistry and Chemistry of Crops. Published by Suri Assoiciates, Coimbatore.

Piper, C.S. 1942. Soil and Plant Analysis, Inter Science Publishers, Inc., New York.

Robert K.M. Hay. 1993. Volatile Oil Crops. Publisher: Longman Science & Technology.

11

Leaf Analysis, Standards and Index Tissues of Different Crops and Interpretation of Leaf Analytical Values

Leaf analysis and its importance

Improved fertilizer management for crops is important in view of today's need to reduce production costs, conserve natural resources, and minimize possible negative environmental impacts. These goals can be achieved through optimum management of the fertilizer applied. Understanding the crop nutrient requirements and using soil testing to predict fertilizer needs are keys to fertilizer management efficiency.

Plant tissue testing is another tool for use in achieving a high degree of precision in fertilizer management. Timely tissue testing can help diagnose suspected nutrient problems or can simply assist in learning more about fertilizer management efficiency. Leaf analysis (also called stem leaf analysis, tissue analysis or foliar analysis) is the most precise method of monitoring plant nutrient levels. While soil analysis reveals the levels of essential soil nutrients, leaf analysis shows the grower exactly what the plant has successfully absorbed. Leaf analysis is especially helpful in detecting nutrient deficiencies before they affect plant health and yield.

Importance

Chemical analysis of plant foliage is an important tool for establishing and maintaining a proper fertilizer programme in soil fertility management especially for fruit plantings.

Leaf analysis can be used to confirm or diagnose a problem associated with a nutrient shortage or excess, and more importantly to prevent the development of a nutrient disorder in crops. It would also reveal that certain fertilizers being used are not necessary and results in the most economical fertilizer programme. Analysis must be properly taken.

In other instances, a series of analyses may be necessary to arrive at a proper explanation.

Paired comparisons, one from normal and one from the abnormal condition, are frequently helpful.

Foliar analyses made over a period of years can indicate an approaching deficiency of a nutrient element before the plant shows any visible symptoms. It is possible then, through proper corrective fertilizer applications, to prevent the deficiency from ever occurring in the crop. By the same token, it is possible to learn when an element may be increasing in a crop toward a level that will reduce crop quality or bring about some other undesirable effect. When this condition is known, steps can be taken to alter the fertilizer programme and cultural practices that influence the uptake of the element from the soil.

Plant tissue analysis and its interpretation

- This, in turn, shows whether soil nutrient supplies are adequate. In addition, plant tissue analysis will detect unseen deficiencies and may confirm visual symptoms of the deficiencies. Toxic levels also may be detected. Though usually used as a diagnostic tool for future correction of nutrient problems, plant tissue analysis from young plants will allow a corrective fertilizer application during the same season. Not all abnormal appearances are due to a deficiency. Some may be due to too much of certain elements. Also, symptoms of one deficiency may look like those of another. A plant tissue analysis can pinpoint the cause, if it is nutritional.

A plant analysis is of little value

- If the plants come from fields that are infested with weeds, insects, and disease organisms;
- If the plants are stressed for moisture; or
- If the plants have some mechanical injury.

The most important use of plant analysis is as a monitoring tool for determining the adequacy of current fertilization practices. Sampling a crop periodically during the season or once each year provides a record of its nutrient content that can be used through the growing season or from year to year. With soil test information and a plant analysis report, a producer can closely tailor fertilization practices to specific soil-plant needs. It also may be possible to prevent nutrient stress in a crop if the plant analysis indicates a potential problem developing early in the season. Corrective measures can be applied during the season or, if the crop is perennial, during the next year. Combined with data from a soil analysis, a tissue analysis is an important tool in determining nutrient requirements of a crop.

Critical concentrations

- As reported in the section on nutrient deficiency symptoms, there is a general concentration range for each essential element that results in normal plant growth. This is called the adequate or sufficient nutritional concentration range (Fig. 1). Plant growth remains relatively constant within the range of concentrations found in the zone of sufficiency.
- The so-called critical concentration occurs at the point where growth is reduced 10% because of a shortage of the element in question. The critical concentration is in the transition zone, which is the borderline between elemental sufficiency and deficiency. Critical concentrations for an element can be different depending on stage of growth and plant part used for the reference tissue.
- The zone of sufficiency (level part of the graph) is the area where an increase in tissue nutrient concentration is not accompanied by an increase in growth (Fig. 1). This is the range in nutrient concentrations in which the grower should attempt to control the fertilizer program. The objective is to maintain tissue nutrient concentrations on the lower side of the range with good fertilization techniques. Managing plant nutrient concentrations on the right of the zone indicates over fertilization and resulting luxury consumption of nutrients by the plant.
- The deficient zone occurs at tissue elemental concentrations lower than those in the transition zone and is accompanied by a drastic restriction in growth. Plants show deficiency symptoms as the nutrient concentration falls within this zone. This is the vertical portion of the curve (Fig. 1).
- At the other end of the scale is the toxicity zone where tissue elemental concentrations are greater than those in the adequate zone. A gradual decrease in plant growth occurs in the toxicity zone. As the tissue concentration rises further, toxicity symptoms, often necrosis, begins (Fig. 1).
- The curve shown in (Fig. 1) is obtained by growing plants at a wide range of concentrations of the element being studied. Meanwhile, other nutrients and factors influencing growth are held constant so that changes in growth can be attributed solely to the nutrient being studied. Either greenhouse or field experiments may be designed to generate the data necessary to develop the relationship between plant growth and tissue concentrations of a particular element.

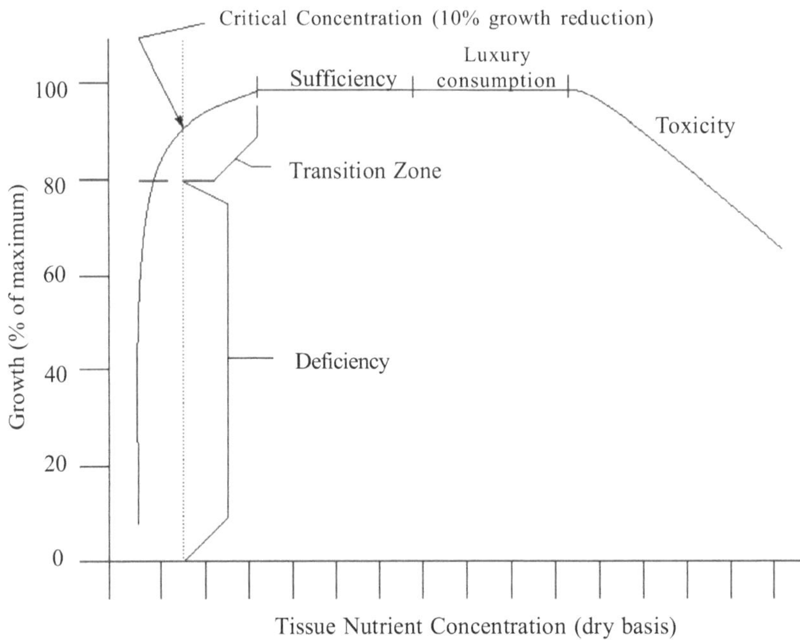

Fig. 11.1 Crop growth in relation to concentration of a nutrient in the diagnostic tissue sample.

- Plant analysis assists in diagnosing nutritional problems or potential problems in the crop from which the samples are taken, *i.e.*, the current crop. Potential problems can be circumvented, particularly if they are discovered early in the crop (before bloom) cycle by routine leaf analyses.
- For example, young cabbage plants that appear normal might have a very low N concentrations for that stage of growth. When checking fertilizer application records, it is found that an error was made, and only 1/10 of the intended rate was applied. Additional N can be applied and the crop can be saved, whereas if symptoms of N deficiency had developed before diagnosis, the crop may have been lost or there may have been a substantial yield reduction. With micro-irrigated or fertigated (drip) crops, the nutritional status of the crop can be monitored continuously, and fertigation adjustments can be made as needed.
- Plant analysis results also have application for fertilizer management of the same crop grown in subsequent seasons. Fertilizer rates can be increased or decreased based on tissue test results and yields of previous crops. Given certain conditions, plant analysis results can be used to manage timing of supplemental sidedress or topdress fertilizer applications.

- Results of plant tissue analysis along with results of soil analysis provide useful tools for the grower in managing the rate and timing of fertilizer applications for vegetables. However, each has limitations and they should not be used for purposes not intended.
- Tissue testing is not recommended if the crop has received foliar sprays containing nutrients, especially micronutrients. There is no way to completely remove residues from leaf surfaces and these residues result in higher test results than actually in the plant tissue.

Diagnostic tissue testing

- By its very nature, diagnostic tissue testing is only undertaken after a problem has been recognized. Often, the grower will see some visual clue that the crop is not as it should be. At this point, information to help make a diagnosis is needed, one component of which may be tissue analysis. Other information, such as soil testing, climatic data, pesticide, and fertilizer records, will often be needed besides nutrient status of the crop before the problem is correctly identified.
- All of the considerations discussed with respect to nutrient monitoring pertaining to laboratory selection and location apply equally to diagnostic sampling. However, sample turnaround time may be the most important, since prompt reaction to some nutrient deficiencies is needed to avoid loss of yield and/or fruit quality.
- Interpretations and recommendations of diagnostic samples should be a two-step process. The first interpretation should be based solely on the concentrations of nutrients found in the tissue sample. In short, do the nutrient levels represent deficiency or toxicity? The information in this circular can help with the answer to this question.
- Secondly, results of samples from the affected area should be compared with those taken from an unaffected area: socalled "normal" and "disorder" areas. The samples should be taken at the same time so that a valid comparison can be made. The distance between the two composite samples should also be as small as possible.
- This comparison will greatly aid in proper diagnosis. Often, a nutrient may be found to be at the lower level of the sufficiency range in the "disorder" sample, immediately making that nutrient suspect. However, comparison between "normal" and "disorder" levels may reveal that the nutrient is of similar magnitude in both samples, indicating that the symptoms may be caused by other factors.

Interpretation

- The sufficiency ranges are given for plant tissues of the crops at ages (or stages in the crop cycle) that research has found appropriate for sampling.
- The analysis data can be used as a guide for attaining improved crop quality and yield.
- For long-term orchard crops, plant tissue nutrient levels can be logged over time and used as a diagnostic tool to assist in developing a fertilizer programme.
- If a tissue level of a nutrient is below the lower end of the sufficiency range, the nutrient should be considered deficient, whereas if the level is above the upper end of the range, the nutrient can be considered as approaching a toxic level.
- The midpoint of the sufficiency range is the target to aim at.
- As the level approaches the lower limit, the nutrient should be added.
- As the level approaches the upper limit, additions of the nutrient should be withheld. It is important to be near the midpoint for most nutrients, because imbalances in the ratios of nutrients can affect crop growth.
- Because environment plays a major role in nutrient uptake and crop development, the sufficiency ranges given here should be considered as general guides.
- In addition to variation due to environment effects, different crop cultivars may have different critical levels.

Recommended plant tissue nutrient levels: Sufficiency values and ranges are presented in the following Table.

Table: Recommended plant tissue nutrient levels: sufficiency values and ranges*

Vegetable	N (%)	P (%)	K (%)	Ca (%)	Mg (%)	S (%)	B (ppm)	Fe (ppm)	Mn (ppm)
Beans, snap 10 most recently matured leaves	3.00-6.00	0.25-0.75	1.80-4.00	0.80-3.00	0.25-1.00	0.23	20-75	50-400	30-300
Bitter melon 3 similar to cucumber	NA	NA	NA	NA	NA	NA	NA	NA	NA
Broccoli, heading most recently matured leaf	3.20-5.50	0.30-0.75	2.00-4.00	1.00-2.50	0.23-0.75	0.30-0.75	30-100	70-300	25-200
Cabbage, Chinese (won-bok) (mature plant) 1st fully developed leaf	3.00-4.00	0.40-0.70	4.50-7.50	1.90-6.00	0.20-0.70	0.40-0.80	26-100	40-300	25-200
Cabbage, head (2-3 months) wrapper leaves	3.60-5.00	0.33-0.75	3.00-5.00	1.10-3.00	0.40-0.75	0.30-0.75	25-75	30-200	25-200
Cabbage, mustard (pak-choi) fully expanded leaf	2.39-5.51	0.36-0.80	2.86-5.74	1.29-3.21	0.19-0.35	0.41-0.77	19-39	85-363	35-52
Carrots (mid-grown) most recently matured leaf	2.10-3.50	0.20-0.50	2.50-4.30	1.40-3.00	0.30-3.00	0.20-0.40	30-100	50-300	60-300
Cauliflower, Heading most recently	3.30-4.50	0.33-0.80	2.60-4.20	2.00-3.50	0.24-0.50	NA	30-100	30-200	25-250

Contd.

matured leaf									
Celery (6 wks) most recently	1.60-2.00	0.30-0.60	8.60-10.00	2.20-3.50	0.25-0.50	NA	25-50	30-100	10-100
matured petiole									
Corn, sweet (5-6 wks)	3.50-4.50	0.30-0.50	2.80-3.80	0.50-0.90	0.20-0.50	0.21-0.70	8-25	50-300	31-300
5thmature leaf from tip									
Corn, sweet (early tassel) 5th	2.70-3.50	0.30-0.50	2.50-3.50	0.70-1.00	0.20-0.50	0.21-0.70	8-25	50-350	31-300
mature leaf from tip									
Cucumber (bud to small fruit)	4.50-6.00	0.34-1.25	3.90-500	1.40-3.50	0.30-1.00	0.04-0.70	25-60	50-300	50-300
5th mature leaf fromtip									
Daikon similar to radish	NA	NA	NA	NA	NA	NA	NA	NA	NA
Dasheenmost recently	2.69	0.44	2.92	2.29	0.45	0.43	34	47	50
matured leaf									
Eggplant (bud to small fruit)	4.00-5.00	0.30-0.60	3.50-5.00	1.00-2.50	0.30-1.00	NA	25-75	50-300	40-250
most recently matured leaf									
Ginger root (2-3 months)	3.00-3.50	0.24-0.33	3.90-5.70	1.10-1.30	0.50-0.80	0.35-0.40	80-112	110-160	125-250
3rd leaf blade									
Lettuce, head (mature)	3.80-5.00	0.45-0.60	6.60-9.00	1.50-2.25	‘0.50- 0.80	NA	23-50	50-100	25-250
wrapper leaves									
Lettuce, Romaine (mature)	3.50-4.50	0.45-0.80	5.50-6.20	2.00-2.80	0.60-0.80	NA	25-60	40-100	11-250
plant wrapper leaves									
Onion, dry (1/3- ½ grown)	5.00-6.00	0.35-0.50	4.00-5.50	1.00-3.50	0.30-0.50	0.50-1.00	22-60	60-300	50-250
top ½ of leaf, no white portion									
Peas, Chinese (1st bloom)	4.00-5.00	0.30-0.80	2.00-3.50	1.20-2.00	0.30-0.70	NA	5-60	50-300	25-400
most recently developed leaflet									

Contd.

Pepper, bell (1st fruit filled) most recently matured leaf	4.00-6.00	0.35-1.00	4.00-6.00	1.00-2.50	0.30-1.00	NA	25-75	60-300	50-250
Potato, white (30 cm tall) most recently maturedleaf	4.00-6.00	0.20-0.50	4.00-11.50	0.60-1.00	0.50-1.50	0.19-0.36	25-50	50-150	30-450
Potato, white (tubers ½ grown) most recently matured leaf	3.00-4.00	0.25-0.40	6.00-8.00	1.50-2.50	0.70-1.00	NA	40-70	40-100	30-250
Potato (sweet potato) (1/2 grown) most recently matured leaf	3.30-4.50	0.23-0.50	3.10-4.50	0.70-1.20	0.35-1.00	NA	25-75	40-100	40-250
Pumpkinmost recently matured leaf	4.00-6.00	0.30-0.50	3.00-5.00	1.20-2.50	0.30-1.00	NA	25-75	50-200	50-250
Radish (mid-season) most recently matured leaf	3.00-5.00	0.30-0.70	4.00-7.50	3.00-4.50	0.50-4.50	NA	25-125	50-200	50-250
Squash, zucchini (mature, no fruiting plant) most recently matured leaf	4.90	0.93	3.71	1.40	0.49	0.32	38	112	82
Taro, lowland most recently matured leaf	3.00-4.50	0.30-0.50	3.00-5.50	0.75-1.50	0.25-0.50	0.20-0.30	10-25	125-150	300-400
Taro, dryland most recently matured leaf	4.20-4.50	0.33-0.35	3.70-4.20	0.90-1.50	0.36-0.43	NA	24-26	150-200	280-315
Tomato (mid-bloom to 1st cluster) leaves opposite or below top flower cluster	3.50- 5.00	0.70- 1.30	6.00-10.0	1.40- 2.20	0.30-0.70 NA	NA NA	25-75 NA	60- 300	50- 250
Watercress most recently matured compound leaf	4.20-6.00	0.70-1.30	4.00-8.00	1.00-2.00	0.25-0.50	NA	25-50	50-100	50-250

Contd.

Fruits									
Atemoya most recently matured leaf	2.50-3.00	0.16-0.20	1.00-1.50	0.60-1.00	0.35-0.50	NA	15-40	NA	NA
Avocado (5-7 months)									
Most recently matured nonfruiting terminal leaf	1.60-2.00	0.08-0.25	0.75-2.00	1.00-3.00	0.25-0.80	0.20-0.60	50-100	50-200	30-500
Banana (Production fields) mid-section leaf strips	3.50-4.50	0.20-0.40	3.80-5.00	0.80-1.50	0.25-0.80	0.20-0.80	10-50	75-300	100-1000
Cantaloupe (Small fruit to harvest) 5th leaf from tip	4.09-5.00	0.25-0.80	3.59-5.00	2.30-3.20	0.35-0.80	0.23-1.40	25-60	50-300	‘50-250
Cheimoya most recently matured leaf	2.25-3.10	0.15-0.25	1.00-2.00	0.55-1.25	0.30-0.50	NA	15-60	NA	NA
Coffee (maturity to flower set) 4th leaf pairfrom tip	2.30-3.00	0.12-0.20	2.00-2.50	1.00-2.50	0.25-0.40	0.10-0.20	40-75	70-125	50-200
Grapefruit (nonfruiting) most recently matured leaf	2.40-3.00	0.15-0.50	0.80-2.20	1.50-5.50	0.25-0.75	0.15-0.05	30-100	60-200	25-200
Guava (production trees) 3rd leaf pair from tip on fruiting shoots	1.25-1.70	0.15-0.20	1.25-1.75	0.80-1.75	0.25-0.50	NA	NA	NA	NA
Honeydew melon (small fruit to harvest) 5th leaf from tip	4.09-5.00	0.25-0.60	3.59-4.50	2.59-3.20	0.35-0.80	0.23-1.20	25-60	50-300	50-250
Lime (production fields, nonfruiting) most recently matured leaf	2.40-3.00	0.15-0.50	1.60-2.50	1.50-5.00	0.25-1.00	0.15-0.50	30-100	60-200	20-200

Lychee (fruit setting) most recent matured leaf behind spring flush	1.50-1.80	0.14-0.22	0.70- 1.10	0.60-1.00	0.30-0.50	1.10-0.16	25-60	50-100	100-250
Macadamia (mature orchards) mature leaves from mew growth	1.45-2.00	0.08-0.11	0.45-0.65	0.65-1.00	0.08-1.10	0.24	40-100	50	50-1500
Mango (after flowering) most recent fully matured leaf	1.00-1.50	0.13-0.18	0.30-1.20	2.00-3.50	0.20-0.40	0.06-0.22	30-100	70-100	60-200
Orange, navel (nonfruiting) most recently matured leaf	2.40-2.70	0.12-0.16	0.70-1.10	1.50-2.60	0.25-0.70	0.20-0.40	30-100	60-120	25-200
Papaya petiole under most recently set fruit	1.20-1.35	0.17-0.21	2.70-3.40	1.00-3.00	0.40-1.20	NA	20-50	25-100	20-150
Persimmon (production trees) recently matured leaf from mew growth	1.75-2.50	0.10-0.25	2.25-4.50	1.25-3.30	0.18-0.50	0.20-0.45	45-100	50-150	200-1000
Pineapple (inflorescence start) first fully expanded leaf, basal white section	1.50-2.50	0.10-0.30	3.00-6.50	0.40-1.20	0.30-0.60	0.10-0.30	30-75	75-200	50-400
Tangerine (production trees, nonfruiting) most recent fully matured leaf	3.00-3.40	0.15-0.25	0.90-1.10	NA	0.17-0.44	NA	31-100	NA	NA
Watermelon (1st flower to small fruit) 5th leaf from tip, omit unfurled leaf	4.00-5.50	0.30-0.80	4.00-5.00	1.70-3.00	0.50-0.80	NA	25-60	50-300	50-250

Data Not Available

References

Garn, A. Wallace. 2010. Plant Tissue Analysis Complements Soil Test. www.wlabs.com copyright
http://www.avocadosource.com/books/TraynorJoe1980/IDEAS_PG_79-84.pdf
Illustrated Guide to Sampling for Plant Analysis. 2009. Spectrum Analytic Inc
Piper, C.S. 1942. Soil and Plant Analysis, Inter Science Publishers, Inc., New York
www.bettersoils.com/planttestingart.pdf
www.servitechlabs.com/Portals/0/PlantSampling1.pdf

12

Rapid Tissue Tests for Soil and Plant

The crop growth and productivity is conditioned by many factors of which, the nutrient status (Content) of plant parts such as leaf, stem, etc play a critical role. Moreover the leaf and stem are considered as the indicator parts of plants for assessing the nutrients content of plant. Each crop plant requires the essential element at a specific concentration at different growth stages and it is known as 'critical level'. When the nutrient content of plant depletes below the critical level the plants may exhibit some symptoms. The requirement or otherwise the availability of nutrients can be assessed by i) plant diagnosis ii) soil analysis and iii) plant analysis by two methods a) by qualitative test and b) by quantitative estimation. Based on the plant or soil tests, the required nutrients can be applied for crops to sustain the growth and rectify the deficiency disorders. The rapid tissue test would pave way for rectifying the nutritional problems for quick recovery, however the quantitative estimation of both plant and soil for nutrients concentration will be more useful and economic for applying fertilizers either as basal or foliar and would be the long term strategy to cope up with nutritional problems.

For rapid tissue test to assess the nutrient status, different parts of plant should be taken as indicator tissue and some of the representative crops are furnished below:

Crops	Nutrient					
	N	P	K	Ca	Mg	S
Cereals	Stem/Midrib	Leaf blade	Leaf blade	Leaf lamina	Leaf lamina	Leaf blade
Pulses	Petiole	Leaf blade	Leaf blade	Leaf lamina	Leaf lamina	Leaf blade
Oil seeds	Petiole	Leaf blade	Leaf blade	Leaf lamina	Leaf lamina	Leaf blade
Cotton	Petiole	Petiole	Petiole	Petiole	Petiole	Petiole
Banana	Leaf lamina	Leaf lamina	Leaf lamina	Leaf lamina	Leaf lamina	Leaf lamina
Papaya	Petiole	Petiole	Petiole	Petiole	Petiole	Petiole
Vegetables	Petiole, Leaf blade	Petiole Leaf blade	Petiole, Leaf blade	Petiole, Leaf blade	Petiole, Leaf blade	Petiole, Leaf blade

Fruit trees

Either leaf blade/mid rib/leaf lamina can be taken.

Ornamentals, Tea, Coffee, *etc.*,

The leaf blade should be taken.

Micronutrients

The leaf lamina/ leaf blade/ mid rib portion of leaf can be taken.

Procedure for tissue test

1. Nitrogen

Reagent: 1-% diphenylamine in conc. sulphuric acid.

Small bits of leaf or petiole are taken in a petridish and a drop of 1% diphenylamine is added. The development of blue colour indicated the presence of nitrate – nitrogen. The degree of colouration indicates the amount of nitrogen present in that leaf.

Dark blue : Sufficient nitrogen

Light blue : Slightly deficient nitrogen

No colour : Highly deficient nitrogen

2. Phosphorous

Reagents: (1) Ammonium molybdate solution, (2) Stannous chloride powder.

Eight gm ammonium molybdate is dissolved in 100 ml of distilled water. To this, add 126 ml of conc. Hydrochloric acid (Hcl) and volume is made up to 300 ml with distilled water. This stock solution is kept in an amber coloured bottle and at the time of use it is taken and diluted in the ratio of 1:4 using distilled water.

A tea spoonful of freshly chapped leaf bits are taken in a test tube and 10 ml of ammonium molybdate reagent is added and kept for few minutes. After shaking, a pinch of stannous chloride is added. Colour development is observed.

Dark blue : Sufficient phosphorus

Bluish green : Slightly deficient phosphorus

No colour : Highly deficient phosphorus

3. Potassium

Reagent: (1) Sodium cobalt nitrate reagent, (2) Ethyl alcohol (95%).

Take 5 gm cobalt nitrate and mix with 30 gm of sodium nitrate in 80ml of distilled water. To this, 5ml of glacial acetic acid is added. The volume is made up to 100 ml distilled water. Dilute reagent prepared (5 ml) with 15 mg sodium nitrate to 100 ml using distilled water.

Finally cut leaf bits are taken in a test tube and 10 ml diluted reagent is added and shaken vigorously for few a minutes and kept for 5 minutes. Then add 5 ml of ethyl alcohol reagent, allowed to stand for 3 minutes. The solution is observed for the formation of turbidity.

No turbidity : Deficiency of potassium

Slightly turbidity : Moderate deficiency

High turbidity : Sufficient potassium

4. Calcium

Morgan's Reagent: 30 ml of glacial acetic acid and 100 grams of sodium acetate are dissolved in a little of distilled water

Procedure: 0.5 g of finally cut plant material is taken into a glass vial (both of healthy plant and deficient plant in different vials) and 5 ml of Morgan's reagent is added in test tube. After allowing it to stand for 15 minutes, 2 ml of glycerin and 5 ml of 10% ammonium oxalate is added and the solution is shaken for 2 minutes. The turbidity resembling after 15 minutes indicate the amounts of calcium in normal plant tissue.

5. Magnesium

Reagents

(1) 5% pure sucrose solution

(2) 2% Hydroxylamine hydrochloride

(3) Titan yellow

(4) Sodium hydroxide

150 mg of Titan yellow is dissolved in 75 ml of 95% ethyl alcohol and 25 ml distilled water. This solution is stored in darkness.

Procedure

To a tea spoonful of finely cut material, following reagents are added in sequence. One ml of 5% sucrose solution, 1 ml of 2% Hydroxylamine hydrochloride and 1 ml of Titan Yellow. Finally solution was made alkaline with 2 ml of 10% NaOH. Red colour indicates the presence of magnesium and yellow colour indicates absence or traces of Magnesium.

6. Iron

Finely cut leaf materials (0.5g) are taken into a glass vial and 1ml of con. HCl is added in it. After 15 minutes, 10ml of distilled water and 2-3 drops of con HNO_3 are added. 10 ml of this solution is pipetted out into a specimen tube after 2 minutes and 5ml of 20% ammonium thiocynate is added and stirred. Further, 2 ml of amyl alcohol is added, shaked well and allowed to stand for few minutes. The intensity of red colour in amyl alcohol layer indicates the quantity of iron.

Plant-Sap Quick Test for Nutrient Analysis (Developments in USA)

- Much of the diagnostic information deals with analysis of dried plant material (whole leaves, leaf blades, or petioles). The time period from sampling to recommendations for problem correction can be excessive for many situations involving deficiencies. Cost of routine sampling and analysis that involves many samples might be too high for many growers. However, the cost of tissue testing should be compared to the crop value at stake. Costs are often cited as hindrances to routine use of tissue testing in a fertilizer management program. Growers like the idea of tissue testing but may be reluctant to use it in a routine and timely fashion.
- An alternative, for certain nutrients, to traditional laboratory analysis is a nutrient determination made on the fresh plant sap. Procedures for plant sap analysis have been available for years, but recently the techniques have been improved to make them more accurate and easier to use in the field. Most of these in-field plant sap "quick tests" should be used in conjunction with periodic laboratory analysis done on dried whole leaves.
- Plant sap analysis kits are available in a range of sophistication from simple, hand-held "colorimeters" and ion-specific electrodes to sophisticated portable laboratory units that can test for a multitude of nutrients and chemicals. Plant sap kits can test for several plant nutrients but the user needs to evaluate the need for speed versus accuracy for the nutrients to be determined. For example, a sap test kit may not have the desired accuracy for certain micronutrients compared to traditional laboratory analyses using whole leaves.
- Currently, plant sap test kits appear to have most utility for the mobile nutrients such as N, P, and K. These elements, particularly N and K, make up the bulk of nutrients applied as fertilizers to vegetable crops and also are the ones most often managed during the growing season, which makes plant sap testing particularly attractive for these elements. A good example is N management through the season with micro-irrigation. The routine use of a calibrated plant sap quick test could help a micro-irrigation

manager make decisions regarding N scheduling for the crop. Proper management of N could reduce the overall fertilizer applications to that crop.

- The kits, described below, have been adapted to determine nitrate and K concentrations of fresh plant sap from petioles of most-recently-matured leaves. The initial work was conducted for tomato, although some work also has been done for other crops (cantaloupe, broccoli, cucumber, squash, and collards).
- Plant sap test kits are easy to use and result in rapid evaluations of plant sap for nitrate and potassium.
- For sap testing, petioles collected from MRML are used for analyses. Most-recently-matured leaves (MRML) are leaves that have essentially ceased to expand and have turned from a juvenile light-green color to a darker-green color. A random sample of a minimum of 25 petioles should be collected from each "management unit" or "irrigation zone." Management units larger than 20 acres should be subdivided into 20-acre blocks. Leaves with obvious defects or with diseases should be avoided. Sampling should be done on a uniform basis for time of day (best between 10 AM and 2 PM), and for interval after rainfall or fertilization.
- For tomatoes, the sample is usually the fifth or sixth leaf from the tip. Whole leaves are collected from the plant and the leaf blade tissue and leaflets are then stripped from the petiole. For tomatoes, a petiole of six to eight inches in length remains. Petioles are chopped into about one-half inch segments. If analysis is not to be conducted immediately in the field, then whole petioles should be packed with ice and analyzed within a few hours of collecting. Given more extreme environmental field conditions (high temperature and bright sun), more dependable results are obtained by making measurement in the lab or office than outdoors.
- Chopped petiole pieces are mixed and a random subsample (about 1/4 cup) is crushed in a garlic press, lemon press, or hydraulic press (obtainable from HACH Co.). Expressed sap is collected in a small beaker or juice glass and stirred.
- Early in the season, when sap nitrate-N concentrations are high, the sap might need to be diluted. Dilution makes it possible to read the nitrate-N levels within the scales of some test kits. Dilution also will minimize the interference of the green chlorophyll color of the sap on the reading of colorimetric testing systems. Some users have reported success with charcoal-filtered sap. This procedure is particularly good for dark sap that does not need to be diluted. Slightly different results will be obtained

with filtered and unfiltered sap and users should standardize procedures with one method. With tomatoes, a dilution of 50 or 60 parts deionized or distilled water to one part sap is needed. Later in the season, a dilution of 20 to 1 will usually suffice. Diluting can be accomplished by using a laboratory pipette and graduated cylinder or less precisely, with an eyedropper. The pipette method is recommended for highest accuracy. Diluted sap is stirred completely prior to use in the test kits.

- For the Quant strip test, a test strip is removed from the container (keep strips cool when not in use) and dipped for a second into the diluted sap. Following 60 seconds, the pink or purple color developed on the test pad on the end of the strip is compared to the calibrated color chart provided with the kit. Interpolation will be needed for readings between any two color blocks on the chart. An alternative is to use a newly developed strip color reader. This reflectometer provides for more quantitative evaluation of the color on the strip. Readings are made in parts per million (ppm) nitrates which can be converted into ppm nitrate-N by dividing by 4.45.
- For the HACH colorimeter, two viewing tubes are filled with diluted sap. One tube is placed in its slot in the "comparator." Contents of one powder reagent pillow are emptied into the second diluted sap sample and the tube mixed for one minute. After mixing, the tube is placed in its slot in the "comparator" and left for one minute. After one minute, the colors in the viewing slots are matched by rotating the color wheel, and the resulting ppm of nitrate-N read from the dial.
- For the Cardy meters, plant sap is pressed from the petioles and a drop is placed on the Cardy meter, covering both electrode spots on the meter. The meter must be calibrated with standard ion solutions before measuring ion concentration in the sap and again between every 6 or 8 measurements. There are specific meters for nitrate-N and K.
- The results are interpreted as to sufficiency ranges.

13

Quality of Irrigation Water

Introduction

It is necessary to ascertain the quality of irrigation water so as to gauge the possible effects of this water on the soil. The irrigation water often results in water logging, salinity and alkalinity problems. It is therefore very important to know its quality. Following are the basic criteria or characteristics on which irrigation water quality is judged.

1. Presence of total soluble salts in irrigation water.
2. Proportion of carbonates and bicarbonates in relation to calcium and magnesium in irrigation water.
3. Proportion of sodium in relation to other cations in irrigation water.
4. Concentration of specific ions in irrigation water.

Several parameters are employed to judge the quality of irrigation waters No parameter in complete in judging the quality of irrigation water. Further the quality of water should be judged by considering the crop and soil factors. Any judgment without considering the above factors may not be useful. Researchers devised many parameters by following above said criteria. Some of the commonly adopted parameters used to judge the quality of irrigation water are:

Electrical conductivity of irrigation water

- Potential salinity
- Puri's salt index
- Sodium adsorption ratio
- Soluble sodium percentage
- Residual sodium carbonate
- Permeability index
- Boron concentration in water

Parameters used to judge the quality of irrigation waters

a. Electrical conductivity (EC)

It is a measure of total soluble salts in water. Since there exist proportionality between concentration of salts in water and electrical conductance the EC is commonly employed as one of the important parameter to judge water quality with respect to salinity. There are many classification available based on EC namely USDA, USSR and Tamil Nadu STL etc. But most widely used one is USDA classification.

b. Potential salinity (PS)

This considers the concentrations of chloride and sulphate in irrigation water. The potential salinity of irrigation water can be derived using the formula

$PS = ½\ SO^{2-}_4 + Cl^-$

c. Puri's salt index (PSI)

Puri's salt index accounts relationship between concentrations of Na and Ca ions in a given irrigation water. PSI can be arrived by applying the concentration values of water in the following formula

$PSI = (Total\ Na^+ - 24.5) - (Total\ Ca^{2+} - Ca\ as\ CaCO_3) \times 4.85$

d. Sodium adsorption ration (SAR)

This explains the relationship between concentrations of sodium and calcium & magnesium of irrigation water. SAR of given water can be worked out by using the formula

Sodium Adsorption Ratio (SAR) =Na / {0.5(Ca + Mg)}^0.5

e. Soluble sodium percentage (SSP)

The proportion of sodium with that of total cations in a given irrigation water is considered to judge the quality under soluble sodium percentage. in relation to total cations in water

$$SSP = \frac{Na}{\text{Total Cations } (Ca + Mg + K + Na)} \times 100$$

f. Residual sodium carbonate (RSC)

In this parameter the proportion of carbonates & bicarbonates in relation calcium & magnesium in irrigation water is considered. Applying concentration values of carbonate, bicarbonate, calcium and magnesium in the following formula the RSC of water is arrived

$RSC = (CO_3^{2-} + HCO_3^-) - (Ca + Mg)$

g. Permeability Index (PI)

This parameter explains ratio between Na & HCO_3 and cations like calcium, magnesium and sodium in irrigation water.

Permeability Index (PI) = $Na+HCO_3 \times 100/Ca+Mg+Na$

h. Boron concentration

The specific ion effect is taken here to judge the quality of irrigation water. Any ion for that matter exceeds certain limit considered poor quality and may not be suitable for irrigation. Boron being highly soluble in water the concentration of boron is often found to be high in many locations. Hence B concentration in water is commonly considered as one of the parameter to judge the quality of irrigation water.

A part from boron, excess concentrations of ions like carbonate, bicarbonate, chloride, sulphate, nitrate, fluorine and heavy metals including nickel, lead, cadmium *etc.*, are also injurious to growing plants in specific locations. Boron is excessively available in some locations which makes the water quality poor. Further, the critical limits are varying with ions, soil type and crops.

References

Agriculture Handbook 60. U.S. Dept. of Agriculture. Washington D.C

Chhabra, R. 2009. Soil Salinity and Water Quality. Published by Taylor and Francis.

http://www.p2pays.org/ref/20/19718.htm

USSL, 1954. Diagnosis and Improvement of Saline and alkali Soils. USDA Agr. Handbook No. 60., Washington, D.C

Wilcox, L.W., Magistad, O.C. 1943. Interpretation of Irrigation Water Quality and relative Salt Tolerance of Crops. U.S. Bureau of Plant Industry, Washington D.C., USA.

www.fao.org/docrep/t0667e/t0667e08.htm

14

Management of Poor Quality Irrigation Water in Crop Management

Introduction

In classification of irrigation water, it is assumed that the water will be used for average conditions with respect to soil texture, infiltration rate, drainage, quantity of water used, climate and salt tolerance of the crop. Large deviations from the average for one or more of these variables may make it unsafe to use what, under average conditions, would be a good water; or may make it safe to use what, under average conditions, would be a water of doubtful quality. The relationship to average conditions must be kept in mind in connection with the use of any general method for the classification of irrigation water.

However, the quality of irrigation water is evaluated, mainly by the quantity of dissolved salts, proportion of anions, monovalent and divalent cations, sodium adsorption ratio, residual sodium carbonate, potential salinity, specific ion toxicity, proportion of magnesium, calcium in water etc.

1. Salinity based on E.C – USDA system

< 250μ.mhos/cm : Low saline : Good

250 – 750μ.mhos/cm : Medium saline : Fair

750 – 2250μ.mhos/cm : High saline : Poor

2250μ. mhos/cm : Very high saline : Very poor

2. Potential salinity (PS)

PS of water	**Soil type suitable**
1 and 3 me/L	fine, medium and coarse textured soils
3 and 15 me/L	safely used in medium and coarse textured soils
15 and 20 me/L	safely used only in coarse textured soils

3. Puri's salt index (PSI)

PSI of water	Suitability
negative	good quality and suitable for irrigation
positive	poor quality water and unsuitable for irrigation

4. Sodium adsorption ratio (SAR)

SAR of water	Suitability
<10	Low sodic water
10.1 – 18.0	Medium sodic water
18.1 – 26.0	High sodic water
> 26	Very high sodic water

5. Soluble sodium percentage (SSP)

SSP of water	suitablility
< 60 per cent	good quality and suitable for irrigation

6. Residual sodium carbonate (RSC)

RSC of water	Suitablility
<1.25 me/L	good quality water
1.25 – 2.5 me/L	fair quality water
>2.5 me/L	poor quality water and unsuitable for irrigation

7. Permeability index (PI)

$$PI = \frac{Na + HCO_3}{Ca + Mg + Na} \times 100$$

- PI <60 per cent – good quality water and suitable for irrigation
- PI >60 per cent – poor quality water and unsuitable for irrigation

8. Boron concentration

< 0.5 ppm - Low

0.5 – 1.0 ppm - Medium

1.0 – 2.0 ppm - High

2.0 – 4.0 ppm - Very high

> 4.0 ppm – Excessive

9. Fluorine concentration

< 1 ppm - Good

> 1 ppm - Problematic

10. Combined evaluation on the basis of EC & SAR

United States Salinity Laboratory (Richards, 1954) published a semi logarithmic diagram, which rates water with respect to EC and SAR. Four classes of each salinity and SAR were divided as shown below in the table.

Salinity Hazard	Class	EC m.mhos/cm	Sodium Hazard	Class	SAR
Low	C1	< 0.250	Low	S1	< 10
Medium	C2	0.250 - 0.750	Medium	S2	10 – 18
High	C3	0.750 - 2.250	High	S3	18 – 26
Very high	C4	>2.250	Very high	S4	>26

Management of poor quality irrigation waters

Saline water

1. Crop management

Growing crops that are tolerant high level soil salinity *e.g.*: Cotton, ragi, barley, sugar beat, beet root, curry leaf, bermuda grass, saline grass, spinach etc.

Crops that are tolerant to soil salinity at medium level are paddy, wheat, onion, maize, sunflower, castor, grape, pomegranate, tomato, cabbage and potato.

Crops that are tolerant to low level of soil salinity are garden beans, Reddish, lime etc. Black gram and green gram are sensitive to soil salinity. Crops are to be chosen based on the soil salinity level.

Relative tolerance of crops to salinity

Plant species	Threshold salinity (dS m-1)
Field crops	
Cotton	7.7
Sugar beet	7.0
Sorghum	6.8
Wheat	6.0
Soybean	5.0
Groundnut	3.2
Rice	3.0
Maize	1.7
Sugarcane	1.7

Vegetables	
Tomato	2.5
Cabbage	1.8
Potato	1.7
Onion	1.2
Carrot	1.0
Fruits	
Citrus	1.7

2. Soil/cultural management

Growing crops in raised beds will reduce accumulation of salt around root zone. Planting seedlings / sowing seeds on sloppy ridges decreases accumulation of salts around root zone. Mulching soil prevents evaporation which reduces accumulation of salts due to capillary rise of water at surface of soils. Providing drainage in water logged areas also helps to reduce salt accumulation.

3. Fertilizer management

Addition of extra dose of nitrogen to the tune of 20-25% of recommended level will compensate the low availability of N in these soils. Addition of organic manures like, FYM, compost, *etc.* helps in reducing the ill effect of salinity due to release of organic acids produced during decomposition. Green manuring (Sunhemp, Daincha) and / or green leaf manuring also counteracts the effects of salinity.

4. Irrigation management

Proportional mixing of good quality (if available) water with saline water and then using for irrigation reduces effect of salinity. Alternate furrow irrigation favors growth of plant than flooding. Drip and sparkler irrigation systems aim to reduce the use of water which is favorable for growth of plant since slat accumulation also reduced with low usage of water.

All the above four management practices suitably integrated to reduce the soil salinity, which is favorable for better growth of plants and ultimately for better yields. Management of saline soils becomes essentials and unavoidable particularly in areas where both soil as well as irrigation water are saline in nature.

Sodic water

Calculated quantities of gypsum added based on RSC along with irrigation water to reduce the ill effect of sodicity on soil and plants. The techniques employed for soil sodicity can also be followed.

References

Agriculture Handbook 60. U.S. Dept. of Agriculture. Washington D.C.

Ayers, RS. 1977. Quality of Water for Irrigation. Journal of Irrigation and Drainage iv., ASCE, 103:135-154.

Doneen, L.D. 1954. Salinization of Soils by Salt in Irrigation Water. Trans. Amer. Geophy. Union. 35:943-950.

Gordon Johnson and Hailin Zhang.2010. Classificationof Irrigation Water Quality. http://www.soiltesting.okstate.edu/Extn_Pub/F-2401web.pdf

http://cursos.puc.cl/unimit_agl_014/almacen/1219762224_lgurovic_sec4_pos0.pdf

http://www.fao.org/docrep/t0667e/t0667e05.htm

Scofield, C.S. 1935. Salinity of Irrigation Water. Smithsonian Institute, Annual Report, Washington D.C., pp. 275-287.

USSL, 1954. Diagnosis and Improvement of Saline and Alkali Soils. USDA Agr. Handbook No. 60. Washington, D.C.

15

Soil and Water Pollution

I. **Soil pollution** is any physical or chemical change in soil that adversely affects the health of plants and other organisms living in and on it.

a. Soil pollution is important not only in its own right but because so many soil pollutants tend to move into surface water, groundwater, or air.

b. The chief soil pollutants are salts, petroleum products, heavy metals, and agricultural chemicals. Point source releases of industrial solvents (both LNAPLs and DNAPLs) and feedstock chemicals (*e.g.*, for making plastics) often occur as the result of LUSTs (leaking underground storage tanks).

c. Salinization, a common problem in irrigated arid and semiarid regions, makes soil unfit for growing most crops. It is extremely difficult to remove excess salts from salinized soils.

d. A variety of techniques are used for soil remediation, which is cleaning up contaminated soil.

1. These include physical techniques such as

a. Soil vapor extraction (SVE), moving air through the soil.

b. Thermal conduction heating, heating the soil before SVE.

2. Bioremediation methods include

a. Landfarming, fertilizing, liming, and repeatedly plowing contaminated soil to stimulate bacterial decomposition of high molecular weight hydrocarbons like those in diesel oil.

b. Bioventing, pumping air at low rates through soil to provided O_2.

c. Biosparging, pumping air into aquifers to provided O_2.

d. Phytoremediation, particularly useful for removing heavy metals.

II. **Water pollution-** World Health Organisation (WHO) estimates that 21% of communicable diseases is due to water pollution. It consists of any physical or chemical change in water that adversely affects the health of humans and other organisms.

a. Sewage, wastewater carried off by drains or sewers, contributes to enrichment (fertilization of water) and produces an oxygen demand as it is decomposed.
   i. Microorganisms use the process of cell respiration to break down sewage into carbon dioxide, water, and similar materials.
   ii. Biochemical oxygen demand (BOD) is the amount of oxygen needed by microorganisms to decompose sewage and other organic wastes.
   iii. BOD spikes immediately downstream from sewage outfalls killing fish and other O_2requiring organisms. Well over one billion O_2-consuming bacteria can grow on the organic nutrients contained in each ml of raw sewage.
b. Disease-causing agents, such as bacteria, viruses, protozoa, and parasitic worms, are transmitted in sewage. The award for most impressive performance by a water-borne pathogen in the U.S. is currently held by *Cryptosporidium* which gave 370,000 people the trots in the Milwaukee area in 1993.
   i. The common intestinal bacterium *E. coli* is used as an indication of the amount of sewage present in water and as an indirect measure of disease-causing organisms.
   ii. The fecal coliform test determines the presence of *E. coli* in water, or other related bacteria that are indistinguishable from *E. coli* in the test.
   iii. Unless it carries a disease-associated plasmid, *E. coli* is not usually a pathogen. It is, however, easy to detect, so it is used as a "proxy" for the presence of potential fecal pathogens. The drinking water standard requires less than one fecal coliform bacterium per 100 ml of water.
c. Inorganic plant and algal nutrients, such as nitrogen and especially phosphorus, contribute to enrichment.
   i. Fertilizer runoff from agricultural and residential land is a major contributor of inorganic plant and algal nutrients. Livestock produce 20 times as much feces and urine in the U.S. as do people, but no law requires livestock waste to be routed to sewage treatment plants.
   ii. Fertilizer and livestock waste runoff from Midwestern fields is carried by the Mississippi River to the Gulf of Mexico, where it causes an low-oxygen condition known as hypoxia over a large area resulting

in the formation of a "dead zone" the size of New Jersey. Even larger dead zones have been reported in the Baltic and Black Sea.

iii. Nitrate reaching the ocean can actually increase the danger of cholera in nearby coastal areas. The cholera bacterium, *Vibrio cholerae*, is supported by some of the organic compounds exuded by the algae that bloom in response to nitrate fertilization.

d. Many organic pollutants are synthetic and do not decompose readily. Some of these, such as pesticides, solvents, and industrial chemicals, are quite toxic to organisms. These synthetic organics are often nonpolar and actually dissolve better in the lipids within animal tissues than they do in water. Consequently, they are subject to biomagnification and increase in concentration as they pass up the food chain in much the same way that DDT does.

e. Inorganic chemicals, contaminants that contain elements other than carbon, include toxins such as lead and mercury.

i. Small amounts of lead occur naturally in the environment, but most lead contamination can be traced to human activities. Children with low levels of lead in their blood may suffer from partial hearing loss, hyperactivity, attention deficit, lowered IQ, and learning disabilities.

ii. Mercury is used in a variety of industrial processes and is also released during the combustion of coal. Once in a body of water, mercury is converted to methyl mercury, which readily enters the food web and accumulates in fishes. Methyl mercury compounds are highly toxic, and are able to cross the blood-brain barrier. Low levels of mercury "cause neurological problems such as headache, depression, and quarrelsome behavior", such as that exhibited by the Mad Hatter of Alice in Wonderland, who in Lewis Carroll's day would have received high occupational exposure to Hg.

III. **Eutrophication**, the enrichment of oligotrophic lakes or other standing-water ecosystems by nutrients, results in high photosynthetic productivity, which supports an overpopulation of algae.

a. Eutrophic lakes tend to fill in rapidly as dead organisms settle to the bottom. Eutrophication also kills fishes and causes a decline in water quality as large numbers of algae die and decompose rapidly. The "dead zone" in the Gulf of Mexico as merely a marine version of what would be called eutrophication in a lake.

b. Artificial eutrophication, also called cultural eutrophication, is enrichment of a standing-water ecosystem that occurs at an accelerated rate due to human activities such as fertilizer runoff and sewage.

c. Whether natural or artificial, eutrophication ultimately results in the conversion of a fresh standing water body into a wetland.

IV. **Water pollutants** come from both natural sources and human activities.

a. Pollution that enters the water at specific sites, such as pipes from industrial or sewage treatment plants, is called point source pollution.

b. Nonpoint source pollution, also called polluted runoff, comes from the land rather than from a single point of entry. Agriculture is the principal producer of nonpoint source pollution. The terms point and nonpoint source are also applied in some discussions of air pollutants.

c. Three major sources of human-induced water pollution are:

i. Agriculture (72% of total). Fertilizer runoff generates pollution by plant and algal nutrients. Animal waste runoff adds more P and N plus BOD. More than 95% of stream and river water samples and 50% of groundwater samples in the U.S.were contaminated by residues of at least one pesticide.

ii. Municipalities (sewage and urban runoff). Urban runoff includes anything that goes down storm drains: road salt, sediments from construction sites, dog excrement, traffic emissions, oil and grease. Most towns and cities direct water collected in storm drains into the same sewer system used to transport sewage to the treatment plant in a scheme known as a combined sewer system. A heavy rain introduces enough flow into the combined system to overwhelm the treatment capacity of the plant, so the gates are opened and raw, untreated combined sewer overflow is released to the receiving body of water. This is one of the ways in which Boston Harbor got as contaminated as it did.

iii. Industries. The nature of the water pollution generated depends upon the kind of the industry. Locally, there is a felt dyeing company that releases (or at least, released) its dyes to the Connecticut River at night, when the spectacularly colored waste stream is less noticeable.

V. **Pollutants** that seep from sanitary landfills, underground storage tanks, and agricultural operations can contaminate groundwater.

a. Currently, most of the groundwater supplies in the United States are of good quality, although there are some local problems.

b. Because cleanup of polluted groundwater is very costly, takes years, and in some cases is not technically feasible, it is important to prevent groundwater contamination from occurring in the first place.

i. Some of the most commonly occurring yet difficult to clean up examples of groundwater contamination are those involving DNAPLs

(Dense, NonAqueous-Phase Liquids) such as trichloroethylene (TCE), because they sink through groundwater either to pool on the confining layer at the bottom of a contaminated, unconfined aquifer or become trapped within pores at various depths throughout the aquifer.

ii. Hydrocarbon solvents are easier to remove because they float on the water table(They are LNAPLs).

iii. Volatile dissolved organic contaminants can be expelled from groundwater by sparging, i.e., bubbling air through an unconfined aquifer. Once in the soil atmosphere they can be removed through a process called soil vapor extraction (SVE)

iv. Many common chlorinated DNAPL solvents (like TCE) can be dechlorinated to innocuous ethylene gas by anaerobic bacteria, if the aquifer is rendered anoxic, *eg.,* through the addition of ethanol or benoate. Similar anaerobes have at least partially dechlorinated PCBs in Hudson and Housatonic river sediments.

VI. **Wastewater** treatment may include

a. Primary treatment (the physical settling of solid matter). The low level of agitation maintained to allow rapid settling precludes much lowering of dissolved BOD during this stage of treatment. For about 10% of wastewater treatment facilities, primary treatment is the only treatment, before chlorination and discharge. For an on-site sewage disposal system, the septic tank provides this function.

b. Secondary treatment (the biological degradation of organic wastes). In this stage, the focus is on the reduction of BOD and the elimination of the vast majority of bacteria. Some schemes for secondary treatment involve vigorous aeration of the primary effluent. In other schemes, such as that used in Amherst, the primary effluent is repeatedly trickled over a large tankful of rocks that have become coated with bacteria and stalked protozoa that eat suspended bacteria. In either scheme, a tremendous amount of mineralization of nitrogen and phosphorus occurs. Typically, 3-5% of the BOD, 70% of the P, and 50% of the N survives this stage. For 62% of treatment facilities, this is the final stage before chlorination and discharge.

c. Tertiary treatment (the removal of special contaminants such as organic chemicals, nitrogen, and phosphorus).

i Phosphate, the principal promoter of freshwater eutrophication, is removed by flocculation with alum.

ii The engineering to remove nitrate is more difficult because it is best done through biological denitrification, an anaerobic process.

iii Treatments systems for the removal of salts (through electrodialysis) and residual dissolved organics (e.g., w/ activated carbon) are quite expensive to operate.

iv All of the above tertiary treatments are cheaply and naturally provided by a constructed or reclaimed wetland.

d. Primary, secondary, or tertiary effluents must all be chlorinated by law before discharge to eliminate any remaining pathogens. Unfortunately, when dissolved organics are present, chlorination results in the formation of chloroorganic compounds, including several suspected carcinogens, which have proven toxic to many freshwater species.

e. The slimy mixture of bacteria-laden solids that settles out during sewage treatment is called primary sludge (formed during primary treatment) and secondary sludge (formed during secondary treatment).

f. One of the most pressing problems of wastewater treatment is disposal of the sludge that results from primary and secondary treatments. Five common methods of sludge disposal are:

i direct application to farmland as fertilizer

ii incineration

iii ocean dumping (illegal since 1991) 4. burial in a sanitary landfill

iv anaerobic digestion.

g. Digested sludge resembles humus and makes an excellent, soil additive that serves as a slow-release source of nutrients and a promoter of good soil structure. Even metal rich, city sludges may be useable for crops such as turf or woody ornamentals that are not destined for human or livestock consumption.

VII. **Laws attempt to control water pollution.** Monitoring and enforcement are difficult, however.

a. The Safe Drinking Water Act requires the EPA to establish maximum contaminant levels (MCLs, *e g.*, 10 ppm for NO_3-N) for water pollutants that might affect human health. In 1998, 25% of U.S. public water systems reported instances where MCLs were exceeded, but only 6% were in serious violation.

b. The quality of rivers, lakes, aquifers, estuaries, and coastal waters in the United States is most affected by the Clean Water Act, which requires the EPA to establish national emission limitations for wastewater that is discharged into surface waters.

c. Legislation has been more effective in controlling point source pollution than in controlling nonpoint source pollution.

d. The many laws that address groundwater pollution operate in isolation from one another and often at cross-purposes.

Further Readings

Agriculture Handbook 60. U.S. Dept. of Agriculture. Washington D.C.

Bornstein, J., W.E. Hedstrom, and F.R. Scott. 1980. Technical Bulletin on Oxygen Diffusion Rate Relationships in Three Soil Conditions. Life Sciences and Agriculture Experiment Station, University of Marine at Orono.

Brady, N. C.and R. R. Weil; 2002. The Nature and Properties of Soils, 13th. Edition. Kluwer Academic Publishers. Printed in the Netherlands

Daniel Hillel. Fundamentals of Soil Physics. Academic Press.

Das D. K. 2004. Introductory Soil Science 2nd Edi. 2004. Kalyani Publishers

Garn A. Wallace. 2010. Plant Tissue Analysis Complements Soil TEST. www.wlabs.com copyright

Gildayal, B.P. and R.P. Tripathi. 1987. Soil Physics. Published by John Wiley & Sons (Asia) Pte Ltd., Singapore

Hesse, P. R. 1972. A Textbook of Soil Chemical Analysis. Published by Chemical Publishing Company Ltd., New York. pp. 556.

http://eusoils.jrc.ec.europa.eu/soil_sampling/index.html

http://soil.gsfc.nasa.gov/pvg/prop1.htm

http://soils.usda.gov/technical/classification/

http://www.ars.usda.gov/SP2UserFiles/ad_hoc/54450000Glomalin/Soil%20aggregate%20analysis.pdf

http://www.ctahr.hawaii.edu/mauisoil/a_factor_ts.aspx

http://www.ehow.com/list_7387607_list-four-types-soil-structure.html

http://www.epa.gov/oust/cat/mason.pdf

http://www.sci.sdsu.edu/SERG/techniques/mfps.html

http://www.spectrumanalytic.com/services/analysis/plantguide.pdf

Jackson, M.L. 1973. Soil Chemical Analysis. Prentice Hall of India. New Delhi. pp. 498

Kemper, W.D. and W.S. Chepil. 1965. Size Distribution of Aggregates. In: Methods of soil analysis, Part I. Ed: Black, C.A. Agronomy No. 9. American Society of Agronomy 499-509.

Oswal, M.C. 1983. Soil Physics. Vikas Publishing House, New Delhi

Piper, C. S. 1942. Soil and Plant Analysis, Inter Science Publishers, Inc., New York

Rattan Lal, R. Lal, Manoj Shukla. 2004. Principles of Soil Physics. CRC press

Soil Education - http://soils.usda.gov/education/index.html

Soil Quality Test Kit Guide - http://soils.usda.gov/sqi/assessment/test_kit.html

Soil Science Education Home Page - http://soil.gsfc.nasa.gov

T.D. Biswas and S.K. Mukherjee.1995. Text Book of Soil Science. Tata McGraw-Hill Publishing Company Limited, New Delhi.

Other Publications on Horticulture

S.No.	Title	Author	ISBN	Year
1	A Colour Handbook on Practical Plant Pathology	Yadav, Vijay	9789385516177	2015
2	A Colour Handbook: Landscape Gardening	Singh, Alka & B.K. Dhakuk	9789383305889	2015
3	A Handbook of Minerals,Crystals,Rocks and Ores	Alexander, P.O.	9788190723787	2009
4	A Handbook of Soil-Plant-Water-Fertilizer and Manure Analysis	Durai, M.V.	9789381450185	2014
5	A Handbook on Irrigation and Drainage	Panigrahi, Balram	9789381450888	2013
6	A Text Book of Seed Science and Technology	Padmavathi, S.	9789381450437	2012
7	Abiotic Stress and Physiological Process in Plants	A. Bhattacharya	9789385516696	2017
8	Abiotic Stress and Plant Physiology *(in 2 parts)*	A. Bhattacharya and Luxmi Vijaya	9789385516900	2018
9	Abiotic Stress Tolerance in Crop Plants: Breeding and Biotechnology	Roy, Bidhan	9788189422943	2009
10	Acid Soils: Their Chemistry and Management	Sarkar, A.K.	9789381450383	2013
11	Advances and Challenges in Agricultural Extension and Rural Development	Rathakrishnan, T.	9789380235035	2009
12	Advances in Agri-Management	Rakhi Gupta *et al.*	9789385516894	2017
13	Advances in Fruits and Vegetables Processing and Preservation	V.K.Joshi		2018
14	Advances in Nutrient Dynamics in Soil-Plant Systems for Improving Nutrient use Efficiency	Elanchezhian, R.	9789385516962	2017
15	Advances in Preservation and Processing Technologies of Fruits and Vegetables	Rajarathnam, S.	9789380235523	2011
16	Advances in Protected Cultivation	Singh, Brahma, Balraj Singh	9789383305179	2014
17	Advances in Soil Borne Plant Diseases	Naik, Manjunath	9788189422813	2008
18	Agricultural Biotechnology: Indian Print	Persley, G.J.; et al	9788189422097	2006
19	Agricultural Extension: Worldwide Innovations	Saravanan, R.	9788189422967	2008
20	Agricultural Marketing: Perspectives and Potentials	Bhat, Anil & S.P. Singh	9789385516153	2016
21	Agricultural Plant Biochemistry	Nagaraj, G.	9789383305551	2015
22	Agricultural Statistics: A Guide for Competitive Examinations	Kushwaha, K.S.	9789381450314	2012
23	Agriculture Bioinformatics	Keshavachandran, R.	9789383305421	2015
24	Agri-Food Crops: Processing,Value Addition, Packaging and Storage	R.Sasi Kumar	9789381450406	2012
25	Agri-Horticultural Biodiverstiy of Temperate and Cold Arid Regions	Zeerak, Nazir Ahmad	9789381450086	2012
26	Agrobiodiversity and Sustainable Rural Development	Soam, S.K. & M.Balakrishnan	9789383305896	2015
27	Agro-enterprises For Rural Development and Livelihood Security	Sharma J.P. & S.K.Dubey	9789380235851	2011

28	Agroforestry for Increased Production and Livelihood Security	Sushil Kumar Gupta Pankaj Panwar & Rajesh Kaushal	9789385516764	2017
29	Agroforestry Systems for Resource Conservation and Livelihood Security in Lower Himalays	Panwar,P & Dadhwal	9789381450215	2012
30	Agroforestry: Principles and Practices	Patra, Alok Kumar	9789381450765	2013
31	Agroforestry: Systems and Practices	Puri, Sunil & Pankaj Panwar	9788189422622	2007
32	Agroforestry: Systems and Prospects	C.B.Pandey & O.P.Chaturvedi	9789381450970	2014
33	Agro-Informatics	Vanitha, G & M.Kalpana	9789380235714	2011
34	Agronomy: Principles and Practices	E. Somasundaram & M. M. Amanullah	9789385516740	2017
35	Agrotechnology Manual: Including Nursery Management and Practices	Etomes, Marcel N	9789383305018	2014
36	An Introduction to Intellectual Property Rights	Pathak, Manju	9789383305124	2014
37	An Introduction to Nanotechnology	Rathinasamy, A.	9789381450413	2012
38	Ancestral Knowledge in Agri-Allied Science	Saha, Ratan Kumar	9789383305216	2014
39	Applied Computational Biology and Statistics in Biotechnology and Bioinformatics (Set of 2 Vols.)	Roy, A.K.	9789380235929	2012
40	Applied Statistical Techniques	Imam, Ekwal	9789383305537	2015
41	Applied Statistics for Agricultural Sciences	Venkatesan, D.	9789383305285	2014
42	Approaches for Incorporating Drought and Salinity Resistance in Crop Plants	Chopra,V.L. & R.S.Paroda	9789383305742	2015
43	Aromatic Plants: Vol.01. Horticulture Science Series	Skaria, B.P. et.al.	9788189422455	2007
44	Auddaniki ke Adharbhoot Sidhant Tatha Nashijeev Prabandhan	Tiwari, A.K.	9789380235998	2012
45	Bananas and Plantains: Postharvest Management, Storage,Ripening and Processing	Narayana, C.K.	9789383305452	2015
46	Basic Concepts in Statistics	Kushwaha, K.S.	9788189422400	2009
47	Basics of Horticulture: 2nd Revised and Expanded ed. (As per Revised ICAR Syllabus)	Peter, K.V.	9789383305735	2015
48	Basics of Mutation Breeding	Thirugnanakumar, S.	9789383305193	2014
49	Basics of Research Methodology	Imam, Ekwal	9789383305438	2015
50	Biochemical Aspects of Plant Physiology: Technology and Methodology	Bhattacharya, Amitav	9789383305902	2016
51	Biochemistry,Molecular Biology and Biotechnology: Instant Notes	Gajera, H.P.	9789383305520	2015
52	Bioinformatics in Agriculture: Tools and Applications	Balakrishnan, M. & Dam Roy ed.	9789381450925	2014
53	Biometrical Methods in Horticultural Science	Nirmal Sharma, V.K. Wali, Parshant Bakshi	9789385516504	2016
54	Biostatistics: Basic Concepts and Methodology	Kushwaha, K.S.	9789383305469	2014
55	Biotechnology in Horticulture: Methods and Applications	Peter, K.V.	9789381450918	2013
56	Biotechnology in India: Initiatives and Accomplishments	Niladri, Bag	9789385516252	2016
57	Biotechnology: Practical Manual Series Vol 04	Thara, K.M.	9788190851237	2009
58	Bougainvillea	S.K. Datta, R. Jayanthi & T. Janakiram	9789385516702	2017

59	Breeding and Biotechnology of Flowers: Set of 2 Vols. (Set Price)	Singh, A.K.	9789383305612	2014
60	Breeding and Biotechnology of Flowers: Vol 01: Commercial Flowers	Singh, A.K.	9789383305353	2014
61	Breeding and Biotechnology of Flowers: Vol 02: Garden Flowers	Singh, A.K.	9789383305407	2014
62	Breeding and Protection of Vegetables	Rana, M.K.	9789380235493	2011
63	Breeding of Horticultural Crops: Principles and Practices: 2nd Revised & Expanded ed.	Kumar, N.	9789383305773	2015
64	Breeding,Biotechnology and Seed Production of Field Crops	Roy, Bidhan	9789381450680	2013
65	Cashew Production and Processing Technology: Recent Advances	Gajbhiye, R.C. & B.V. Padhiar	9789383305827	2015
66	Climate Change and Agricultural Food Production	Kibria, Golam et.al.	9789381450512	2013
67	Climate Change and Agroforestry: Adaptation, Mitigation and Livelihood Security	C.B.Pandey, M.K.Gaur & R.K.Goyal		2018
68	Climate Change and Food Security	Datta, M.& N.P.Singh	9788189422387	2008
69	Climate Change and Natural Resources Management	Lenka, & Lenka	9789381450673	2013
70	Climate Change and Plantations in the Humid Tropics	GSHLV Prasada Rao and C.S. Gopakumar	9789385516368	2016
71	Climate Change and Sustainable Agriculture	P Suresh Kumar	9789385516726	2017
72	Climate Mitigation and Carbon Finance: Global Initiatives & Challenges	Sahoo, A.K.	9789381450024	2012
73	Climate Resilient Animal Agriculture	GSLHV Prasada Rao		2018
74	Climate Resilient Crops for the Future	Peter, K.V.	9789383305599	2015
75	Climatic Variability: Impacts on Agriculture and Allied Sectors	Datta, M.:Ed.	9789381450949	2014
76	Commercial Crops Technology: Vol.08. Horticulture Science Series	Kurian, A. & K.V.Peter	9788189422523	2007
77	Commercial Horticulture	Patel, N.L. & S.L. Chawla ed.	9789385516238	2016
78	Computers in Agriculture	M. K.Sharma & Anil Bhat	9789385516160	2017
79	Crop Diseases Management: Principles and Practices	Narayanasamy, P.	9789380235677	2011
80	Crop Diseases: Identification,Treatment and Management	Darwin, Henry	9789380235462	2011
81	Developments in Physiology,Biochemistry and Molecular Biology of Plants Vol 01	Bose, Bandana & A.Hemantaranjan: ed.	9788189422028	2005
82	Developments in Physiology,Biochemistry and Molecular Biology of Plants Vol 02	Bose, Bandana & A.Hemantaranjan: ed.	9788189422929	2008
83	Dimensions of Extension Education	Mohapatra, B.P.	9789381450987	2014
84	Diseases of Field Crops and Their Integrated Management	Sanjeev, Kumar	9789385516283	2016
85	Diseases of Horticultural Crops Identification and Management: With Colour Illustrations	Kumar, Sanjeev	9789383305643	2015
86	Diseases of Vegetable Crops and Their Integrated Management:A Colour Handbook	Mishra, R.K.	9789381450499	2013
87	Drip and Sprinkler Irrigation	Biswas, R. Kumar	9789383305766	2015
88	Droughts in Agricultural Production: Monitoring & Management	Rao, G.G.S.N.	9789385516009	2015
89	Drug Discovery and Development: Traditional Medicine and Ethnopharmacology	Patwardhan, Bhushan	9788189422295	2007

90	Drying Technologies for Foods: Fundamentals and Applications	Prabhat K. Nema Arun S.Mujumdar	9789383305841	2015
91	Drying Technologies for Foods: Fundamentals and Applications: Part-II	Prabhat K. Nema	9789385516399	2016
92	Eco Agriculture Revolution	M.H.Mehta		2018
93	Ecologically Based Integrated Pest Management (Set of 2 Vols.)	Abrol,D.P.& U.Shankar	9789380235950	2012
94	Economics,Marketing and Sales of Agricultural Products	Thakur, S. Nath	9789381450659	2013
95	Efficiency Indices for Agriculture Management Research	Devasenapathy, P. & B.Gangwar	9788189422882	2008
96	Elementary Agriculture	Nainwal & Nainwal	9789385516535	2017
97	Elements of Entomology	H.Lewin, Devasahayam	9789381450635	2014
98	Emerging Technologies of the 21st Century	Roy, A.K.	9789383305339	2015
99	Encyclopedia of Agricultural Meteorology	Dhaliwal, L.K. & S.S.Hundal	9789380235547	2011
100	Enhancing Nutrient Use Efficiency	Ramesh, K	9789385516733	2017
101	Entomology: Novel Approaches	Jain,P.C. & M.C.Bhargava	9788189422325	2007
102	Essence of Horticulture	M.S. Patil, A.R. Karale,	9789385516481	2016
103	Essential Oils and their Applications	Das, Kuntal	9789381450741	2013
104	Ethnomedicinal Plants Resource of Orissa Vol 01	Sahoo, A.K.	9789380235752	2011
105	Evaluation and Impact Assessment of Technologies and Developmental Activities in Agriculture,Fisheries and Allied Fields	Roy, A.K.	9789380235400	2011
106	Experimental Biotechnology: Practical Manual Series 06	Dutta, Sunita & Abhijit Dutta	9789380235721	2011
107	Experimental Phytochemical Techniques	Raaman, N.	9789380235943	2012
108	Extension Management in the Information Age Initiatives and Impacts	Philip, H. & T.Rathakrishnan	9789381450543	2013
109	Extension of Technologies: From Labs to Farms	Anandaraja, N.	9788189422837	2008
110	Family Farming and Rural Economic Development	M.L.Choudhary & Aditya	9789383305858	2015
111	Flower Crops: Cultivation and Management	Singh, A.K.	9788189422356	2006
112	Flowering Trees: Vol.12. Horticulture Science Series	Valsalakumari, P.K.	9788189422509	2008
113	Flowers for Trade: Vol.10. Horticulture Science Series	Sheela, V.L.	9788189422516	2008
114	Food and Nutritonal Security by Sustainable Agriculture:	Mishra, Kumar Bijesh	9789383305049	2014
115	Food Engineering and Technology	Sharma, H.K. & Ashutosh Upadhyay	9789383305483	2015
116	Food Process Engineering and Technology	Pare, Aakash & Mandhyan	9789380235431	2011
117	Food Processing Waste Management: Treatment and Utilization Technology	Joshi V.K. & Satish Sharma	9789380235592	2011
118	Food Product Development and Process Innovations (in 2 parts)	H N Mishra		2018
119	Food Science	Bawa, A.S. & O.P.Chauhan	9789381450147	2013
120	Food Science and Technology: Glossary of Preeminence	Dev Raj	9789380235806	2011
121	Forest Seed Science and Management	Shukla, Gopal	9789385516757	2017
122	Forestry Science: Fundamentals and Terms	Sharad Nema	9789385516382	2016

123	Fruit Breeding	Dinesh, M.R.	9789383305513	2015
124	Fruit Crops: Vol.03. Horticulture Science Series	Radha, T. & Lila Mathew	9788189422462	2007
125	Functional Foods	Mishra, H.N. Rajesh Kapur	9789383305988	2016
126	Functional Foods and Nutraceuticals: Sources and their Developmental Techniques	Riar, C.S. & Saxena, D.C. Ed.	9789383305964	2015
127	Fundamentals of Garden Designing: A Colour Encyclopedia	Roy, Rup Kumar	9789381450307	2013
128	Fundamentals of Ornamental Horticulture and Landscape Gardening	Tiwari, A.K.	9789381450079	2012
129	Fundamentals of Vegetable Production	Rana, M.K.	9789380235707	2011
130	Genetic Diversity and Phenotypic Stability in Crop Plants	S. Thirugnanakumar	9789385516955	2018
131	Genetic Resources and Seed Enterprises: Management and Policies: In 2 Parts	Ram, Hari Har & R.Yadava	9788189422653	2007
132	Genetics and Breeding of Flower Crops *(in 2 parts)*	M.S.Patil	9789385516948	2018
133	Genomics and Genetic Engineering	Satya, Pratik	9788189422776	2007
134	Geographic Information System	Gurugnanam, B.	9788190851282	2009
135	Geoinformatics Applications in Agriculture	Singh, Anil Kumar & U.K.Chopra	9788189422233	2007
136	Geospatial Technologies for Natural Resources Mgt.	Soam,S.K. & P.D. Sreekant	9789381450802	2013
137	GIS: Fundamentals,Applications and Implementations	Elangovan, K.	9788189422165	2006
138	Glimpes of Practical's in Extension Education	K. Pradhan		2018
139	Good Management Practices for Horticultural Crops	M.K. Jatav, *et al.*	9789385516641	2016
140	Guava: Evaluation Studies of Most Promosing Cultivars	Dolkar, Disket	9789385516849	2016
141	Heterosis Breeding in Vegetable Crops	Rai, Nagendra	9788189422035	2006
142	Horticulture Science Series Vol 01-12: Set of 12 Vols	Peter, K.V.: Series Ed.	9788189422479	2008
143	ICTs for Agricultural Extension: Global Experiments, Innovations and Experiences	Saravanan, R. ed.	9789380235240	2010
144	ICTs for Transfer of Technologies: Tools and Techniques	Verma, S.R.	9788193014479	2015
145	Identification and Management of Horticultural Pests	Ranjith, A.M.	9789381450567	2013
146	Illustrated Dictionary of Entomology	Paras Nath	9788189422561	2007
147	Illustrated Dictionary of Floriculture and Landscaping	M. Kannan		2018
148	Illustrated Plant Pathology: Basic Concepts	Darwin, Henry	9789380235080	2009
149	Improving Productivity of Drylands by Sustainable Resource Utilisation and Management	Dayal, Devi	9789385516191	2016
150	Indigenous Medicinal Plants and their Practical Utility	Lakshman, H.C.	9789381450116	2012
151	Information Access in Digital Libraries	Stanley, Madan Kumar	9789383305384	2014
152	Information and Communication Technology for Agriculture and Rural Development	Saravanan, R.	9789380235882	2011
153	Information and Knowledge Management: Tools, Techniques and Practices	Roy, A.K.	9789381450628	2013
154	Innovations in Food Processing Technologies	Saiti		2018
155	Innovations in Horticultural Sciences	Peter, K.V.	9789385516344	2016
156	Innovative Horticulture	Arunkumar, K.	9788189422738	2008
157	Integrated Farming System Practices: Challenges and Opportunities	Nanda, Sankarsana	9789385516207	2016
158	Integrated Nutrient Management in Kinnow Mandarin	Bakshi, Manish	9789385516573	2016
159	Integrated Pest Management in the Tropics: In 2 Parts	Abrol, D.P.	9789385516115	2016
160	Intellectual Property Rights Demystified	Ramkumar, Mu.	9788189422875	2008

161	IPR: Drafting,Interpretation of Patent Specifications and Claims	Rathore, N.S.: Ed.	9789381450819	2013
162	Irrigation and Agricultural Drainage Engineering	Biswas, Ranajit Kumar	9789383305247	2015
163	Irrigation Systems Engineering	Panigrahi, Balram	9789380235387	2011
164	Laboratory Manual of Biochemistry: Methods and Techniques	Sengar, R.S.	9789383305025	2014
165	Laboratory Manual of Microbiology: Practical Manual Series: 05	Roy, A.K.	9789380235189	2010
166	Library Services in the Knowledge Web	Veeranjaneyulu & R.Mahapatra	9789381450192	2012
167	Management of Horticultural Crops: Vol.11 Horticulture Science Series: (Part-I & II Combined in 1 Binding)	Pradeepkumar, T.	9788189422493	2008
168	Managing Soil Health for Maximising Crop Productivity	Obeng, F.k, Avornyo,	9789385516788	2017
169	Mechanization of Cultivated Crops	Singh, Surendra	9789383305759	2015
170	Medicinal Plants: Vol.02. Horticulture Science Series	Kurian, A. & A.Sankar	9788189422424	2007
171	Merging Plant Breeding with Crop Biotechnology	Yasin, J.K.:Ed.	9789381450598	2012
172	Methods and Techniques in Plant Physiology	Bhattacharya, A. & Vijay Laxmi	9789383305506	2015
173	Microbial Biotechnology for Sustainable Agriculture, Horticulture and Forestry	Bagyaraj, D.J.	9789380235820	2011
174	Microbial Diversity and Functions	Bagyaraj, D.J. & Tilak, K. et.al.	9789381450109	2012
175	Microbial Diversity and Its Applications	Barbudde, S.B., R.Ramesh & N.P.Singh	9789381450666	2013
176	Mobile Phones for Agricultural Extension	Saravanan, R.	9789383305230	2014
177	Modern Biotechnology and Its Applications (Set of 2 Vols.)	Behera, K.K.	9789381450833	2013
178	Modern Methods in Plant Physiology	Srivastava, Girish Chand	9789380235011	2010
179	Modern Technologies for Sustainable Agriculture	Kumar, Sunil	9789381450611	2013
180	Modern Technology in Vegetable Production	Hazra, Pranab	9789380235325	2011
181	Molecular Biology and Biochemistry	Puttaraju, H.P.	9788189422257	2008
182	Molecular Markers and Plant Biotechnology	Tomar, R.S.	9789380235257	2010
183	Nanotechnology in Agriculture	Subramanian, K.S.	9789383305209	2015
184	Nanotechnology in Soil Science and Plant Nutrition	Adhikari, Tapan	9789381450789	2013
185	Nematology: Fundamentals and Applications	Jonathan, E.I.	9789380235141	2010
186	Novel Food Processing Technologies	Nanda, Vikas & Savita Sharma	9789385516047	2017
187	Numericals and Short Questions in Farm Machinery, Power and Energy in Agriculture	Yadav, Rajvir	9788190723718	2009
188	Nutraceutical Values of Horticultural Products	Dhurendra Singh	9789385516979	2018
189	Nutritional Disorders in Fruit Crops: Diagnosis and Management	Prakash, M.	9789381450956	2013
190	Objective Agriculture	Rupinder Singh	9789385516993	2018
191	Objective Entomology	Devinder Sharma	9789385516290	2018
192	Objective Horticulture	Verma, Anil	9789381450239	2012
193	Oilseeds: Properties,Products,Processing and Procedures	Nagaraj, G.	9788190723756	2009
194	Orchid: Cultivation and Management	Tiwari, A.K.	9789383305001	2014
195	Organic Farming	Singh, A.K.	9789385516139	2015
196	Organic Farming: Scope and Uses of Biofertilizers	Panwar, JDS & Amit Kumar Jain	9789385516184	2016
197	Organic Spices	Parthasarathy, V.A	9788189422844	2008
198	Ornamental Gardening and Landscaping	M. Kannan, *et al.*	9789385516689	2016
199	Ornamental Horticulture	Sindhu,S.S.ed.	9789383305865	2016

200	Ornamental Plants	Sabina, George	9788190851268	2009
201	Pest Management and Residual Analysis in Horticultural Crops	Gulati, Rachna & Beena Kumari	9789381450710	2013
202	Pesticides: Methods of Their Residues Estimation	Kumari, Beena & T.S.Kathpal	9789380235394	2010
203	Pests of Vegetables: Bionomics and Management	Laskar, Nripendra ed.	9789385516016	2015
204	Phal Evam Sabji Parirakshan Digdarshika	Kumari, Arunima	9789380235561	2011
205	Physio-Biochemistry and Biotechnology of Vegetable Crops	Rana, M.K.	9789380235318	2011
206	Physiological Disorders of Fruit Crops	Savreet, Sandhu & Bikramjit Singh Gill	9789381450581	2013
207	Phytochemical Techniques	Raaman, N.	9788189422301	2006
208	Plant Biochemistry: Techniques and Procedures	Nagaraj, G.	9789383305940	2015
209	Plant Diseases: Identification and Management (With Illustrations)	Rai, J.P.	9789383305315	2014
210	Plant Nutrient Disorders: Diagnosis and Management	Sarkar, A.K. & P. Mahapatra ed.	9789385516023	2015
211	Plant Pathogens and Principles of Plant Pathology	Kumar, Sanjeev	9789385516078	2016
212	Plant Secondary Metabolities	Shukla, Y.M. & R.Bhatnagar	9788190851220	2009
213	Plant Taxonomy and Biosystematics: Classical and Modern Methods	Rana, T.S.	9789383305414	2014
214	Plantation Crops	Bhani Ram	9789385516511	2016
215	Plants For Wellness and Vigour	V.L.Chopra	9789386546036	2018
217	Postharvest Management and Processing of Fruits and Vegetables: Instant Notes	Sharma, Satish	9789380235202	2010
218	Postharvest Techniques and Management for Dry Flowers	V.Ponnuswami & Aruna,P.	9789380235868	2011
219	Postharvest Technologies for Commercial Floriculture	Verma, Anil	9789381450048	2012
220	Postharvest Technology and Engineering: An Illustrated Guide	Dev Raj	9789381450451	2012
221	Postharvest Technology and Processing of Horticultural Crops	Pandit, P.S.	9789383305278	2014
222	Postharvest Technology of Horticultural Crops: Practical Manual Series Vol 02	Sharma, Satish & M.C.Nautiyal	9788190851206	2009
223	Postharvest Technology of Horticultural Crops: Vol.07. Horticulture Science Series	Sudheer, K.P. & V.Indira	9788189422431	2007
224	Practical Manual of Entomology (Insects and Non-Insects Pests)	Devasahayam,H.Lewin	9789380235905	2011
225	Practical Manual of Horticulture Crops: Set of 2 Vols.	Verma, Anil Kumar	9789383305711	2015
226	Practical Manual of Horticulture Crops: Vol 01: Production Technologies	Verma, Anil Kumar	9789383305704	2015
227	Practical Manual of Horticulture Crops: Vol 02: Processing and Postharvest Technologies	Vaidya, Devina	9789383305698	2015
228	Practical Plant Biotechnology and Genetics	Rani, Archana	9789383305995	2015
229	Precision Farming in Horticulture	Singh, Jitendar, S.K.Jain,L.K.Dashora	9789381450475	2013
230	Production Technology of Spices,Aromatic,Medicinal and Plantation Crops	Barche, Swati	9789385516061	2016
231	Propagation of Horticultural Crops: Vol 06 Horticulture Science Series	Rajan, S. & B.L.Markose	9788189422486	2007
232	Propagation of Horticultural Plants: Arid and Semi-Arid Regions	Singh, R.S. & R.Bhargava	9789383305254	2014

233	Protected Cultivation of Horticultural Crops	Singh, D.K & K.V.Peter	9789383305155	2014
234	Quality Control for Value Addition in Food Processing	Dev Raj, *et al.*	9789380235578	2011
235	Quantitative Genetics and Crop Breeding	Thirugnanakumar, S.	9789380235981	2012
236	Question Bank in Extension Education	Mohammad, Asif & K. Ponnusamy	9788193014493	2015
237	Question Bank in Forest Science	Gopal Shukla		2017
238	Question Bank in Fruit Science	Jarande, S. D	9789385516795	2017
239	Question Bank in Postharvest Technology	Guleria, SPS & Anil Kumar Verma	9789380235042	2010
240	Recent Advances in Biopesticides	Johri, Jayendra: eds.	9789380235219	2010
241	Recent Trends in Horticultural Biotechnology: In 2Vols	Keshavachandran, R.	9788189422592	2007
242	Rejuvenation in Fruit Trees: Tropical and Subtropical	Kumar, Rajesh	9789385516436	2016
243	Remember Your Humanity: Pathway to Sustainable Food Security	Swaminathan, M.S.	9789381450178	2012
244	Rural Livelihood and Food Security	Wani, M.H.	9789380235936	2012
245	Seed Processing: A Practical Approach	Rakesh C. Mathad	9789385516085	2018
246	Seed Production of Field Crops	Mondal, S.S.	9788190723763	2009
247	Seed Science and Technology	Vanangamudi, K.	9789383305117	2014
248	Seed Testing Techniques for Seed Spices	Sangeeta Yadav & Arun Kumar barholia	9789385516917	2018
249	Soil Conservation: Fully Revised and Updated: 3rd ed.	Hudson, Norman	9789383305971	2015
250	Soil Microbiology and Biochemistry	Hassan, G.Dar	9789380235134	2010
251	Soil Sampling and Methods of Analysis	Pal, Sushant	9789381450574	2013
252	Soil Science: An Elementary Textbook	Puri, A.N.	9789383305063	2015
253	Soil Testing and Analysis: Plant, Water and Pesticide Residues	Brajendra, Patiram	9788189422707	2007
254	Spices,Plantation Crops,Medicinal and Aromatic Plants: A Handbook	Tyagi, S.K.	9788193014486	2015
255	Spices: Vol.05. Horticulture Science Series	Nybe, E.V. and Mini Raj	9788189422448	2007
256	Statistical Designs and Analysis for Agricultural Field Experiments	Katyal, Vijay & D.M.Hegde	9789381450840	2013
257	Statistical Methods for Agricultural Field Experiments	Katyal, Vijay & B.Gangwar	9789380235424	2011
258	Sustainable Agriculture: A Vision for Future	Desai, B.K. and B.T.Pujari	9788189422639	2007
259	Sustainable Environmental Science	Sahu, D.D.	9789381450208	2012
260	Systematics of Fruit Crops	Sharma, Girish	9789380235066	2009
261	Tea: Technological Initiatives	Niladri Bag, Arundhati Bag and L.M.S. Palni	9789385516337	2016
262	Technologies for Sustainable Green Environment	Davamani, V.	9789381450420	2012
263	Temperate Horticulture: Current Scenario	Kishore, D.K.: et. al.	9788189422363	2006
264	Textbook of Floriculture and Landscaping	Singh, A.K.	9789386546005	2017
265	The Basics of Human Civilization: Food, Agriculture and Humanity Vol.01 Present Scenario	Prem Nath	9789381450734	2013
266	The Basics of Human Civilization: Food, Agriculture and Humanity Vol.02 Food	Prem Nath	9789383305377	2014
267	The Chemistry of Soil Processes	Greenland, D.J.	9789383305926	2015
268	The Pomegranate	Hiwale, S.S.	9789380235158	2009
269	The Qur-anic Plants and Animals	Zeerak, Nazir Ahmad	9789383305780	2015

270	The Science of Horticulture Vol 01	Peter, K.V.: eds.	9789380235479	2011
271	The Science of Horticulture Vol 02	Peter, K.V.: eds.	9789380235486	2011
272	The Theory of Sample Surveys and Statistical Decisions	Kushwaha, K.S. & Rajesh Kumar	9788189422899	2009
273	The Weeds of Kumaun Himalayan Region (Uttarakhand)	T.S. Rana & Bhaskar Datt	9789385516474	2016
274	Traditional Agricultural Practices: Applications and Technical Implementations	Rathakrishnan, T.	9789380235028	2009
275	Tuber and Root Crops: Vol.09. Horticulture Science Series	Palaniswami, M.S. & K.V.Peter	9788189422530	2008
276	Turning Plants Into Medicines: Novel Approaches	T, Parimelazhagan	9789381450468	2013
277	Underutilized and Underexploited Horticultural Crops: Vol 01	Peter, K.V.: ed.	9788189422608	2007
278	Underutilized and Underexploited Horticultural Crops: Vol 02	Peter, K.V.: ed.	9788189422691	2007
279	Underutilized and Underexploited Horticultural Crops: Vol 03	Peter, K.V.: ed.	9788189422851	2008
280	Underutilized and Underexploited Horticultural Crops: Vol 04	Peter, K.V.: ed.	9788189422905	2008
281	Underutilized and Underexploited Horticultural Crops: Vol 05	Peter, K.V.: ed.	9789380235288	2010
282	Value Addition in Flowers and Orchids	De, L.C.	9789381450123	2011
283	Vegetable Crops: Genetics Resources and Improvements	Singh,D.K. & H.Choudhary	9789380235509	2012
284	Vegetable Crops: Vol 04. Horticulture Science Series	Gopalakrishnan, T.R.	9788189422417	2007
285	Vegetable Seed Processing	Mathad, Rakesh C. & Basave Gowda	9789385516030	2015
286	Water Management for Enhancing Water Productivity	N.K.Gontia & H D Rank	9789385505728	2018
287	Weed Science	Das, P.C.	9789383305261	2015